STATUS OF FISHERIES IN HARYANA

MRS. SONAL YADAV
PhD Scholar, Department of Zoology,
School of Basic and Applied Sciences, Raffles University,
Neemrana - 301705, Alwar, Rajasthan, India.

MR. ASHOK KUMAR PACHAR
PhD Scholar, Department of Biotechnology,
Chaudhary Devi Lal University,
Sirsa - 125055, Haryana, India.

DR. NAVEEN KUMAR
Assistant Professor, Department of Zoology,
School of Basic and Applied Sciences, Raffles University,
Neemrana - 301705, Alwar, Rajasthan, India.

Title   : Status of Fisheries in Haryana

Author   : Mrs. Sonal Yadav, Mr. Ashok Kumar Pachar, Dr. Naveen Kumar

Edition   : First (July, 2024)

ISBN   : 9788197792700

Published by

Regd. Add.: 254, Khuriyakhatta No. 10, Bindukhatta,
Lalkuan, Nainital - 262402, Uttarakhand, India
Website : www.taneeshapublishers.in
E-mail : taneeshapublishers@gmail.com
Phone : +91 845481 2712, +91 976041 7980

Printed by :

Manipal Technologies Limited, Bengaluru - 560001, Karnataka

PREFACE

The fisheries sector in Haryana has undergone a significant transformation, reflecting a broader narrative of evolution, adaptation, and growth within India's aquatic ecosystems. This book, "Status of Fisheries in Haryana," is an endeavor to document and analyze the myriad aspects of fisheries in this northern state, delving into historical contexts, current practices, and future prospects. It aims to provide a comprehensive understanding of the sector, which is crucial for policy makers, researchers, and practitioners alike.

Haryana, predominantly known for its agricultural prowess, has steadily embraced aquaculture, recognizing its potential to enhance food security, generate employment, and contribute to economic growth. This book covers the biological and ecological aspects of fish species, the economic impact of fisheries, and detailed district-wise production statistics, offering a nuanced view of the sector's dynamics.

In preparing this book, we have meticulously gathered data, insights, and case studies from various sources, ensuring a holistic approach to understanding the fisheries landscape in Haryana. The chapters encompass a wide range of topics, from the taxonomy of fish to the socioeconomic implications of fish farming, highlighting both the challenges and opportunities within the sector. We hope this book serves as a valuable resource for those committed to the sustainable development of fisheries and inspires further research and innovation in this vital field.

Mrs. Sonal Yadav
Mr. Ashok Kumar Pachar
Dr. Naveen Kumar

ABOUT THE BOOK

"Status of Fisheries in Haryana" is an exhaustive exploration of the fisheries sector in Haryana, a state that is progressively making strides in aquaculture and fisheries management. This book delves into the intricate tapestry of fish species, their habitats, and the economic, ecological, and social dimensions of fisheries in Haryana. It serves as a critical resource for policymakers, researchers, academics, and practitioners involved in fisheries and aquaculture.

The book begins with a foundational understanding of fish biology and classification, setting the stage for readers to appreciate the diversity and complexity of aquatic life. It elaborates on the anatomy, physiology, and ecological roles of various fish species, providing a scientific backdrop for the subsequent discussions on fisheries management and aquaculture practices.

In the subsequent chapters, the book offers a detailed overview of the fisheries landscape in India, highlighting regional differences and the significant contributions of different states to the national fisheries output. The focus then narrows to Haryana, presenting a comprehensive analysis of the state's fisheries sector. Historical perspectives trace the evolution from traditional fishing methods to modern aquaculture, illustrating the shifts in practices and the growing importance of fisheries in the state's economy.

A significant portion of the book is dedicated to district-wise fish production in Haryana, presenting in-depth data and comparisons that reveal the geographical variations in fish farming practices and outputs. It highlights key districts like Kurukshetra, Ambala, and Yamunanagar, showcasing their contributions and the factors driving their success in aquaculture.

Economic impact is another critical theme, with discussions on how fisheries contribute to Haryana's GDP, employment, and rural livelihoods. The book emphasizes the sector's role in food security and nutritional well-being, underscoring the importance of sustainable practices to ensure long-term benefits.

Additionally, the book addresses contemporary challenges such as overfishing, habitat degradation, and climate change, proposing strategic interventions for sustainable fisheries management. It explores innovative technologies and methods, such as recirculating aquaculture systems and integrated fish farming, that promise enhanced productivity and environmental sustainability.

"Status of Fisheries in Haryana" is not just a documentation of facts and figures but a comprehensive guide that offers insights into the potential and future of fisheries in Haryana. It is an essential read for anyone interested in the sustainable development of aquaculture and fisheries, providing a roadmap for future research, policymaking, and practice.

TABLE OF CONTENT

CHAPTER 1

INTRODUCTION TO FISHES

1.1 OVERVIEW OF FISHES

Definition and biological classification of fish

Fish, as an aquatic group, have long fascinated humans due to their vast diversity, ecological significance, and resource value. Defining fish and comprehending their biological classification entails an exploration of their distinctive traits, evolutionary journey, and the intricate taxonomic system that categorizes them. This journey unveils a complex and captivating narrative of life in aquatic realms. Fish are primarily delineated by their habitat and physical attributes. They are aquatic vertebrates endowed with gills throughout their lives, fins for locomotion, and scales covering their bodies. Unlike their terrestrial counterparts, fish are specialized for aquatic life, with their physiology and anatomy shaped by this environment. Gills facilitate oxygen extraction from water, essential for their survival, while their streamlined bodies optimize swimming efficiency. Most fish are ectothermic, relying on external sources to regulate body temperature.

The biological classification of fish is deeply embedded in the broader framework of animal taxonomy. Taxonomy, the science of classifying organisms based on evolutionary relationships and shared traits, ranges from kingdom and phylum to genus and species. Fish belong to the kingdom Animalia and the phylum Chordata, encompassing animals with a notochord at some developmental stage. Within Chordata, fish are further classified under the subphylum Vertebrata, indicating the presence of a vertebral column.

Fish are categorized into three major classes: Agnatha (jawless fish),

Chondrichthyes (cartilaginous fish), and Osteichthyes (bony fish), representing significant evolutionary lineages with distinct characteristics and histories.

Agnatha, the most primitive fish group, includes species like lampreys and hagfish. These jawless fish lack true jaws, instead possessing a round, sucker-like mouth adapted for feeding. Lampreys exhibit parasitic behavior, attaching to other fish to feed on blood, while hagfish scavenge dead matter on the ocean floor. The absence of jaws in Agnatha reflects an early stage in vertebrate evolution.

Chondrichthyes, comprising sharks, rays, and skates, possess cartilaginous skeletons rather than bone. This class, divided into Elasmobranchii (sharks and rays) and Holocephali (chimaeras), is known for its well-developed senses, particularly keen smell and electromagnetism detection, crucial for prey location.

Osteichthyes, the largest and most diverse fish class, are characterized by bony skeletons and a swim bladder for buoyancy regulation. Divided into Actinopterygii (ray-finned fish) and Sarcopterygii (lobe-finned fish), they display remarkable morphological diversity, with some members like coelacanths representing ancestors of terrestrial vertebrates.

The evolutionary saga of fish spans millions of years, with significant diversification occurring during the Devonian period, dubbed the "Age of Fishes." This era witnessed the emergence of major fish groups and the transition to tetrapods, paving the way for terrestrial life. Fish continue to evolve and adapt to diverse aquatic environments, from the abyssal depths to freshwater ecosystems. Their remarkable morphological and behavioral diversity reflects their adaptability, with specialized adaptations enabling survival in extreme habitats. Reproductive strategies in fish vary, from oviparity to viviparity and ovoviviparity, each tailored to specific environmental conditions. Parental care is common, with some species exhibiting elaborate behaviors to protect and nurture their offspring.

Ecologically, fish occupy various niches as primary consumers, secondary consumers, and apex predators, playing vital roles in aquatic food webs and nutrient cycling. Human activities, however, pose grave threats to fish populations and their habitats, necessitating urgent conservation measures Efforts such as marine protected areas, sustainable fishing practices, habitat restoration, and pollution control are essential for safeguarding fish biodiversity and ensuring the sustainability of aquatic ecosystems. Understanding fish biology, evolution, and ecology is paramount for addressing these challenges and securing the future of these invaluable aquatic organisms.

The Anatomy and Physiology of Fish: Adaptations for Aquatic Life

The anatomy and physiology of fish reveal a complex array of adaptations that enable these organisms to thrive in diverse aquatic environments. These adaptations encompass various systems, including respiratory, circulatory, muscular, skeletal, digestive, nervous, sensory, reproductive, excretory, endocrine, and immune systems. Fish anatomy is characterized by specialized features that support their life in water, such as gills for respiration, fins for movement, and a streamlined body shape to reduce resistance while swimming. A deeper understanding of these anatomical features provides insights into how fish live, grow, and interact with their environments.

Respiratory System: Fish possess a unique respiratory system designed for extracting oxygen from water, a medium denser than air with lower oxygen content. The primary respiratory organs are the gills, located on either side of the head and protected by the operculum. Gills consist of gill arches that support numerous gill filaments, each covered in tiny structures called lamellae. The lamellae increase the surface area for gas exchange. Water enters the fish's mouth, flows over the gills, and exits through the operculum. Oxygen diffuses from the water into the blood across the lamellae, while carbon dioxide diffuses from the blood into the water. This countercurrent exchange mechanism maximizes the efficiency of oxygen uptake.

Circulatory System: The fish circulatory system is closely linked to their respiratory system. Fish have a closed circulatory system with a heart that pumps blood through the body. The heart typically has two chambers: the atrium and the ventricle. Deoxygenated blood is pumped from the heart to the gills for oxygenation and then circulated throughout the body. This single-loop system ensures that oxygen-rich blood reaches various tissues and organs.

Muscular System: Fish muscles are intricately designed to support locomotion in water. Muscles are organized into segmented bands called myomeres, separated by connective tissue known as myosepta. These segments allow for powerful and coordinated contractions that generate the undulating movements characteristic of fish swimming. The main swimming muscles are located along the sides of the body, with myomeres contracting alternately to create wave-like movements that propel the fish forward. Fins play a crucial role in maneuvering, stability, and propulsion.

Skeletal System: The skeletal system provides structural support and protection for internal organs. Fish skeletons can be either cartilaginous, as in sharks and rays, or bony, as in most other fish species. The vertebral column forms the central axis of the skeleton, composed of vertebrae that house and protect the spinal cord. Bony fish have additional skeletal elements such as the cranium, which encases the brain, and the rib cage, which shields vital organs. The bones of bony fish are lighter yet strong, containing a high percentage of calcium phosphate, providing both rigidity and flexibility.

Digestive System: The digestive system of fish is adapted to their diverse diets. The digestive tract includes the mouth, pharynx, esophagus, stomach, intestines, and accessory organs such as the liver and pancreas. Carnivorous fish typically have large, muscular stomachs that produce strong acids and enzymes to break down protein-rich prey. Herbivorous fish have longer intestines to allow more time for digestion and absorption of plant material. The liver produces bile, aiding in fat digestion, while the pancreas secretes digestive enzymes that further break down food components.

Nervous System: The nervous system includes the central nervous system (CNS), composed of the brain and spinal cord, and the peripheral nervous system, consisting of sensory and motor nerves. The fish brain is highly specialized, with key regions for smell, voluntary movements, sensory processing, vision, and balance. The spinal cord transmits nerve signals between the brain and the rest of the body, while peripheral nerves carry sensory information and motor commands.

Sensory Systems: Fish have an array of sensory systems enabling them to detect environmental changes. The lateral line system allows them to sense vibrations and pressure changes in the water. This system consists of fluid-filled canals along the sides of the body containing sensory cells called neuromasts. Fish also have well-developed vision, with eyes adapted to the aquatic environment, often capable of seeing a wide range of light wavelengths. The olfactory system enables them to detect chemical cues, such as food and pheromones.

Reproductive System: Fish reproduction is highly varied, encompassing a range of strategies and adaptations. Most fish are oviparous, laying eggs that develop outside the mother's body. Fertilization can be external, with males releasing sperm over the eggs. Some fish are viviparous, giving birth to live young that develop within the mother's body. Parental care varies widely, from guarding and fanning eggs to carrying young in specialized brood pouches.

Excretory System: The excretory system maintains homeostasis by regulating water and salt balance and removing metabolic waste products. The kidneys filter blood to remove waste and excess substances. Freshwater fish excrete large amounts of dilute urine to counteract water influx, while marine fish produce concentrated urine to conserve water and excrete excess salts through their gills.

Endocrine System: The endocrine system comprises glands that produce hormones regulating growth, metabolism, reproduction, and stress responses. Key endocrine glands include the pituitary gland, thyroid gland, and adrenal glands. The pituitary gland secretes hormones influencing other endocrine glands, while the thyroid gland regulates metabolism and growth. The adrenal glands produce hormones involved in stress responses and osmoregulation.

1.2 TYPES OF FISHES

Classification by habitat (freshwater, marine, and brackish water fish)

Understanding the diverse ecosystems of fish species hinges on their classification by habitat, primarily freshwater, marine, and brackish water. Each habitat presents distinct environmental conditions, shaping the behavior, physiology, and distribution of fish species.

Freshwater Habitats: Rivers, Lakes, Ponds, and Streams

Freshwater fish inhabit environments with salinity levels typically less than 0.5 parts per thousand (ppt). These habitats include rivers, lakes, ponds, and streams, characterized by relatively stable conditions compared to marine environments. Freshwater ecosystems are diverse, supporting a vast array of species adapted to various niches. In rivers, fish often display adaptations to swift currents, such as streamlined bodies and powerful fins for navigation. Lakes and ponds offer habitats for both surface-dwelling and bottom-dwelling fish, with temperature, oxygen levels, and food availability influencing species distribution.

Freshwater fish play crucial ecological roles, controlling insect populations and serving as prey for larger predators like birds and mammals. For instance, trout are often found in cold, fast-flowing streams, where their streamlined bodies and powerful tails aid in navigating strong currents. In contrast, catfish, which inhabit slower-moving waters, possess barbels-whisker-like sensory organs-to detect food in murky conditions. Many freshwater fish, such as salmon, exhibit anadromous behavior, migrating from the ocean to freshwater rivers to spawn. This migration involves remarkable navigational abilities and physiological changes, allowing the fish to transition between saltwater and freshwater environments. The high productivity of freshwater ecosystems supports a vast array of plant and animal life. Fish regulate the populations of aquatic invertebrates and contribute to nutrient cycling by excreting waste products that fertilize aquatic plants. Additionally, freshwater fish are vital to humans, providing food and recreational opportunities, which support local economies.

Marine Habitats: The Vast Oceans

Marine fish inhabit the world's oceans, covering over 70% of the Earth's surface. These environments range from shallow coastal zones to deep-sea trenches, presenting diverse conditions. Marine fish have evolved various adaptations to thrive in these different habitats. For example, tuna and marlin, which live in the open ocean, have streamlined bodies and powerful muscles for high-speed swimming over long distances, essential for finding food and avoiding predators. Coral reefs, one of the most biodiverse marine habitats, host myriad fish species, including clownfish, parrotfish, and groupers. Clownfish have a mutualistic relationship with sea anemones, gaining protection from predators while providing nutrients to the anemone. Parrotfish feed on algae growing on coral, playing a crucial role in maintaining coral health by preventing algae overgrowth.

Deep-sea environments pose challenges such as high pressure, low temperatures, and darkness. Fish like anglerfish and lanternfish have evolved remarkable adaptations to survive. Anglerfish use bioluminescent lures to attract prey, while lanternfish use photophores to communicate and navigate in the dark. These adaptations highlight the incredible diversity of life strategies in marine habitats. Marine fish are integral to ocean ecosystems, serving as prey for larger predators and as predators themselves, controlling smaller fish and invertebrate populations. They contribute to nutrient cycling and energy flow within the ocean. Many marine fish species are economically important, providing food and livelihoods through commercial and recreational fishing.

Brackish Water Habitats: Estuaries, Mangroves, and Coastal Lagoons

Brackish water fish thrive in environments where freshwater mixes with saltwater, such as estuaries, mangroves, and coastal lagoons. These habitats have fluctuating salinity levels, driven by tides, rainfall, and river discharge. Brackish water fish have specialized osmoregulatory systems to maintain salt and water balance, using cells in their gills, kidneys, and intestines to regulate ion and water uptake and excretion. Estuaries, among the most productive ecosystems, provide breeding and nursery grounds for many fish species. High productivity results from the mixing of nutrient-rich freshwater with seawater, supporting abundant plant and animal life. Fish in these habitats often migrate between different parts of the estuary or between estuaries and the open ocean to exploit resources and breeding sites.

Mangroves, with their complex root systems, offer shelter and breeding grounds for juvenile fish, protecting them from predators and strong currents. Species like mudskippers and mangrove snappers are well-adapted to the dynamic conditions of mangroves. Mudskippers can breathe air and move on land, while mangrove snappers use the root systems for protection and hunting grounds. Brackish water fish play essential roles in their ecosystems, contributing to biodiversity and ecosystem functioning. They connect freshwater and marine ecosystems by moving between these habitats, transporting nutrients and energy across the ecosystem. These fish are also important for local fisheries, providing food and income for communities near estuaries and mangroves.

Classification by taxonomy (cartilaginous vs. bony fish)

Fish, an astonishingly diverse group of aquatic vertebrates, are primarily classified into two taxonomic categories: cartilaginous fish (Chondrichthyes) and bony fish (Osteichthyes). This classification hinges on the composition of their skeletons, with cartilaginous fish possessing skeletons made of cartilage and bony fish having skeletons primarily composed of bone. These two groups encompass a vast array of species, each exhibiting unique anatomical, physiological, and ecological traits. Cartilaginous fish, or Chondrichthyes, are divided into two subclasses: Elasmobranchii, which includes sharks, rays, and skates, and Holocephali, which includes chimaeras. The defining characteristic of cartilaginous fish is their cartilaginous skeleton, which is lighter and more flexible than bone. This structural feature provides several advantages, such as enhanced buoyancy and flexibility, crucial for survival in aquatic environments.

Elasmobranchii, specifically sharks, exhibit a range of adaptations that establish them as proficient predators. Their streamlined bodies and powerful tails facilitate efficient swimming. Shark skin is covered with placoid scales, or dermal denticles, reducing drag and providing protection. Sharks also have multiple rows of replaceable teeth tailored to their dietary needs, ranging from slicing flesh to crushing shells. Rays and skates, also within Elasmobranchii, have flattened bodies and enlarged pectoral fins fused to their heads, enabling them to glide through water or along the ocean floor. Rays often have venomous spines on their tails for defense, while skates lack these spines and are generally less aggressive. Both groups feed on benthic organisms such as crustaceans and mollusks, utilizing their ventrally positioned mouths.

Chimaeras, or ghost sharks, the only extant members of Holocephali, differ significantly from other cartilaginous fish. They possess a single gill opening covered by a flap and a long, tapering body ending in a whip-like tail. Typically found in deep-sea environments, chimaeras mainly feed on invertebrates and small fish. In contrast, bony fish, or Osteichthyes, are divided into two subclasses: Sarcopterygii (lobe-finned fish) and Actinopterygii (ray-finned fish). Bony fish are characterized by their bony skeletons, providing a sturdy framework for muscle attachment and organ protection. Their bodies are covered with overlapping scales-cycloid, ctenoid, or ganoid, depending on the species-allowing for smooth movement through water.

Lobe-finned fish (Sarcopterygii), including coelacanths and lungfish, have fleshy, lobed fins supported by a single bone, resembling terrestrial vertebrate limbs. This anatomical feature signifies an evolutionary link between fish and tetrapods. Coelacanths, deep-sea dwellers with unique hinged skulls, consume larger prey by widening their mouths. Lungfish, adapted to freshwater environments, possess both gills and lungs, enabling them to survive in oxygen-poor water. Ray-finned fish (Actinopterygii) represent the largest and most diverse group of vertebrates, comprising over 30,000 species. Characterized by fin rays, bony spines supporting their fins, they exhibit a wide range of forms, behaviors, and habitats-from tiny seahorses to massive sturgeons. Actinopterygii inhabit diverse aquatic environments, including freshwater rivers and lakes, coastal estuaries, and the deep ocean.

Within Actinopterygii, significant infraclasses include Chondrostei and Neopterygii. Chondrostei includes primitive species like sturgeons and paddlefish, retaining ancestral features such as cartilaginous elements and heterocercal tails. Sturgeons are noted for their elongated bodies and bony scutes, while paddlefish, with their paddle-shaped snouts, detect prey through electrical fields. Neopterygii, the more advanced infraclass, includes the majority of modern bony fish. The division Teleostei, comprising most extant species, is known for highly mobile jaws, reduced bony elements, and symmetrical homocercal tails. Teleosts display varied adaptations to exploit different ecological niches, from the bioluminescent lure of the anglerfish to the specialized teeth of parrotfish for scraping algae off coral reefs.

Bony fish exhibit several adaptations enhancing survival and reproductive success. Many have swim bladders, gas-filled sacs regulating buoyancy. This adaptation is critical for maintaining position in the water column. Some deep-sea fish, however, have lost or modified their swim bladders to withstand high-pressure environments. Reproductive strategies among bony fish are diverse. Many exhibit external fertilization, releasing eggs and sperm into the water. Some species, like salmon, undertake long migrations to natal streams for spawning, displaying remarkable homing abilities. Parental care varies widely, from extensive care to no involvement at all.

Bony fish possess advanced sensory systems for environmental detection and interaction. Their lateral line system senses water movements and vibrations, aiding in navigation, prey location, and predator avoidance. Vision, adapted to various light conditions, and a highly developed olfactory system for detecting chemical signals are crucial for survival. Hearing is facilitated by the inner ear with otoliths detecting sound waves, and some species have specialized structures like the Weberian apparatus, enhancing auditory capabilities. Electroreception is present in some species, aiding in navigation and prey detection in low-visibility environments. The evolutionary success of bony fish lies in their anatomical and physiological versatility, allowing adaptation to diverse environmental conditions. Their robust skeletal structure and advanced sensory systems enable effective environmental interaction. Diverse reproductive strategies and parental care behaviors further enhance their ability to colonize various habitats.

Comparing cartilaginous and bony fish reveals distinct evolutionary pathways. Cartilaginous fish, with specialized adaptations, focus on efficient predation and survival in specific environments, while bony fish exhibit broader forms and behaviors, exploiting diverse ecological niches. Their evolutionary success underscores their ecological significance in aquatic environments.

1.3 EDIBLE AND NONEDIBLE FISHES

Common edible fish species

Edible fish species represent a diverse array of aquatic organisms that play integral roles in global diets, economies, and cultures. Their significance transcends mere sustenance; they contribute to culinary traditions and are emblematic of the ecological interconnectedness of marine ecosystems. A closer examination of these species reveals not only their unique characteristics and culinary versatility but also their ecological importance and the imperative of sustainable management. Salmon stands out as one of the most esteemed and widely consumed fish, cherished for its rich flavor and exceptional nutritional value. Native to the frigid waters of the North Atlantic and Pacific Oceans, salmon undertakes remarkable migrations from ocean to freshwater rivers for spawning, a journey that shapes its distinctive taste and texture. Whether wild-caught or farmed, salmon is esteemed for its abundance of omega-3 fatty acids, promoting cardiovascular health. Its adaptability to various cooking methods further enhances its culinary appeal.

Cod, another dietary staple, boasts a mild flavor and delicate texture, rendering it a preferred choice in European and North American cuisines. Inhabiting the cold, deep waters of the North Atlantic, cod-comprising Atlantic and Pacific species-embodies a harmonious balance of taste and nutritional content. Efforts to ensure the sustainability of cod populations underscore its significance in commercial fisheries, with stringent management measures in place.

Tuna, particularly bluefin, yellowfin, and albacore, commands culinary admiration for its taste and nutritional benefits, yet faces challenges due to overfishing. From the prized bluefin, savored in sushi for its buttery texture, to the versatile yellowfin and canned albacore, tuna offers a spectrum of culinary possibilities. However, conservation efforts are imperative to safeguard tuna populations for future generations.

Tilapia, has risen in prominence due to its mild flavor, affordability, and adaptability to various farming conditions. Originating in Africa, tilapia's global cultivation underscores its significance in aquaculture. While lower in omega-3 fatty acids compared to other species, tilapia remains a valuable source of protein and essential nutrients, contributing to its popularity in diverse culinary applications.

Mackerel, celebrated for its rich flavor and high nutritional content, occupies a prominent place in traditional diets, particularly in Japan, the Mediterranean, and Nordic regions. Its oily flesh renders it a prime source of omega-3 fatty acids and essential vitamins, although careful handling is necessary due to its perishable nature.

Sardines, characterized by robust flavor and nutritional density, are revered in various cuisines, particularly the Mediterranean. Packed with omega-3 fatty acids, vitamin B12, and vitamin D, sardines offer a powerhouse of nutrition whether fresh, smoked, or canned.

Haddock, akin to cod in its delicate flavor and firm texture, is cherished in both commercial and recreational fisheries. Whether enjoyed in traditional dishes or smoked for delicacy, haddock exemplifies the culinary diversity of white-fleshed fish.

Halibut, prized for its mild flavor and substantial texture, represents a culinary delicacy in fine dining establishments. Its versatility in cooking methods underscores its appeal to discerning palates.

Trout, particularly rainbow trout, embodies the essence of freshwater delicacies, offering a mild, nutty flavor and tender texture. Whether baked, grilled, or smoked, trout captivates with its versatility and nutritional benefits.

Sea bass, including European and Chilean varieties, is esteemed for its delicate flavor and flaky texture, lending itself to various culinary preparations. Whether grilled whole or served in high-end restaurants, sea bass exemplifies culinary excellence.

Snapper, known for its firm texture and mild flavor, is a versatile fish esteemed in cuisines worldwide. Whether grilled, baked, or featured in ceviche, snapper offers a delightful culinary experience.

Pollock, a significant player in the commercial fishing industry, boasts a mild flavor and flaky texture, making it ideal for a variety of culinary creations.

Catfish, a Southern favorite, charms with its mild sweetness and firm texture, rendering it ideal for frying, baking, or grilling.

Barramundi, esteemed for its sweet, buttery flavor and meaty texture, represents a culinary treasure in Australia and Southeast Asia.

Sole, revered for its delicate flavor and flaky flesh, epitomizes the elegance of flatfish cuisine, whether Dover or lemon sole.

Pike, celebrated for its firm texture and slightly sweet flavor, captivates both sport fishermen and culinary enthusiasts alike.

Swordfish, prized for its dense flesh and mild flavor, delights in various culinary preparations, particularly grilling and broiling.

Flounder, with its delicate flavor and tender flesh, offers a versatile canvas for culinary creativity.

Anchovies, celebrated for their strong flavor and nutritional richness, add depth to countless dishes across diverse cuisines.

Herring, esteemed for its taste and health benefits, enriches Northern European diets with its omega-3 fatty acids and vital nutrients.

Grouper, cherished for its firm texture and mild flavor, lends itself to a myriad of culinary creations, from grilling to frying.

Hake, valued for its mild flavor and flaky texture, represents a culinary cornerstone in European cuisines.

Monkfish, despite its unconventional appearance, captivates with its firm, meaty texture and mild flavor, gracing tables in diverse preparations.

These common edible fish species symbolize not only culinary diversity but also the delicate balance of marine ecosystems. Appreciating their significance entails not only savoring their flavors but also ensuring their sustainable management for future generations to enjoy.

Nonedible fish species and their ecological roles

Non-edible fish species are essential components of aquatic ecosystems, contributing significantly to biodiversity, nutrient cycling, and habitat maintenance. Despite not being targeted for human consumption due to various factors such as taste, toxicity, or size, these fish play pivotal roles in ecosystem functioning and serve as indicators of environmental health. Understanding the ecological roles of non-edible fish entails exploring their contributions to food webs, interactions with other organisms, and impacts on ecosystem processes.

Cleaner fish, exemplified by wrasses and certain gobies, are crucial for maintaining the health of other fish populations through mutualistic relationships. By removing parasites and dead skin from larger fish, they not only benefit the host but also derive sustenance from this cleaning behavior, thus enhancing the overall health and biodiversity of reef ecosystems.

Scavenger fish, including certain catfish and eel species, contribute to decomposition and nutrient cycling by consuming dead organisms and organic detritus. This process helps prevent disease spread and supports primary productivity and the growth of aquatic plants and algae.

Detritivorous fish, such as loaches and some minnow species, play significant roles in the detrital food web by breaking down organic matter and releasing nutrients back into the ecosystem. They also contribute to maintaining water quality and reducing turbidity.

Unique adaptations, like those seen in mudskippers, allow certain non-edible fish to thrive in specific ecological niches, such as intertidal zones. Their activities aid in sediment aeration, nutrient cycling, and the promotion of benthic organism growth.

Non-edible fish species, including mosquitofish, contribute to controlling the populations of invasive species and pests, thus indirectly benefiting human health and ecosystem management. Predatory non-edible fish act as apex or mesopredators, regulating smaller fish and invertebrate populations, which in turn maintains a balance within the food web and supports biodiversity. Certain non-edible fish species serve as important prey for various predators, thereby facilitating energy transfer through the food web and maintaining ecosystem structure and function. Non-edible fish also serve as bioindicators of environmental health, with their populations reflecting changes in water quality, pollution levels, and habitat degradation. Conservation efforts aimed at protecting non-edible fish contribute to maintaining genetic diversity, resilience, and the overall stability of aquatic ecosystems.

Habitat formation and maintenance by non-edible fish, such as engineer gobies and certain cichlids, enhance habitat complexity, increase available niches, and support a greater diversity of species. Migration of non-edible fish species between habitats contributes to ecosystem connectivity, nutrient transfer, and biodiversity support. Non-edible fish influence sediment dynamics and geomorphology through their behaviors, contributing to habitat heterogeneity and the formation of microhabitats. Their interactions with other aquatic species, through competition or facilitation, underscore their importance in ecosystem functioning. Non-edible fish species have cultural, scientific, and educational value, providing insights into ecology, evolution, and environmental stewardship. They play integral roles in conservation and restoration efforts, with habitat restoration, pollution control, and sustainable water management being key strategies. Threats faced by non-edible fish include habitat loss, pollution, climate change, invasive species, and overexploitation, highlighting the need for conservation measures to protect these vital components of aquatic ecosystems.

1.4 FRESHWATER FISHES

Common freshwater fish species

Freshwater fish species exhibit remarkable diversity, populating water bodies worldwide, from rivers and lakes to streams and ponds, adapting to various ecological niches and environmental conditions. Among the well-known and widely distributed species are the common carp (*Cyprinus carpio*), rainbow trout (*Oncorhynchus mykiss*), largemouth bass (*Micropterus salmoides*), bluegill (*Lepomis macrochirus*), catfish (such as *Ictalurus punctatus*), northern pike (*Esox lucius*), and tilapia (including

Oreochromis niloticus). Each species boasts unique biological and ecological characteristics, rendering them intriguing subjects of study and indispensable components of their ecosystems.

The common carp, originating from Asia, has achieved global distribution due to its popularity in aquaculture and recreational fishing. Exhibiting adaptability to various freshwater habitats, these omnivorous fish possess robust physiques and distinctive barbels. However, their feeding habits can disrupt substrates, impacting water clarity and potentially harming other aquatic species. Rainbow trout, native to North America, are renowned for their vibrant coloration and recreational appeal. Thriving primarily in cold, clear rivers and streams, they exhibit anadromous behavior in some populations, migrating between freshwater and the ocean. Largemouth bass, native to the eastern United States, dominate warm, vegetated waters, preying upon a diverse range of organisms and significantly influencing ecosystem dynamics.

Bluegill, characterized by their colorful appearance, inhabit various freshwater habitats across North America, contributing significantly to the food web as prey for larger species. Catfish, exemplified by the channel catfish, play crucial roles in nutrient cycling within freshwater ecosystems, thriving in diverse environments with their omnivorous diets. Northern pike, apex predators in many freshwater ecosystems, regulate fish populations and are prized by anglers for their size and fighting prowess. Tilapia, introduced globally for aquaculture, possess rapid growth rates and reproductive capacities but can disrupt native ecosystems when introduced to non-native habitats. Beyond these familiar species, freshwater ecosystems harbor a myriad of other fish, each with distinct adaptations and ecological significance. Brook trout, indicators of healthy freshwater systems, prefer cold, clean, fast-flowing waters. Muskellunge, similar to northern pike but larger, maintain balanced fish communities in their habitats.

Guppies, native to northeastern South America, exhibit vibrant colors and prolific breeding habits, contributing to their rapid population growth in suitable environments. Freshwater eels, like the American eel, undertake remarkable migrations between freshwater and the ocean, playing essential roles in nutrient cycling. Discus fish, native to the Amazon River basin, are prized in the aquarium trade for their vibrant coloration and complex breeding behaviors. Freshwater stingrays, found in Southeast Asia and South America, glide along riverbeds, consuming a diet of fish, crustaceans, and mollusks. African cichlids, particularly those from the Great Rift Lakes, represent a diverse and extensively studied group of freshwater fish, contributing to the rich tapestry of life in freshwater ecosystems worldwide.

Habitat and distribution

Understanding the habitat preferences and distribution patterns of fish species is paramount in comprehending their ecology, behavior, and evolutionary adaptations. Fish are ubiquitous inhabitants of diverse aquatic environments, spanning oceans, seas, rivers, lakes, streams, estuaries, and wetlands, each offering distinct conditions and resources crucial for their survival and proliferation. This comprehension is indispensable for effective conservation and management strategies, as well as for evaluating the ramifications of environmental transformations and human interventions on aquatic ecosystems.

Oceanic habitats, constituting approximately 71% of the Earth's surface, present a dynamic and formidable milieu for fish species. Adapted to factors such as temperature, salinity, currents, and depth, oceanic fish species exhibit specialization within pelagic, demersal, and benthic zones. Pelagic species, including tuna, mackerel, and herring, navigate the vast open waters, sustaining themselves on plankton and other microorganisms, while exhibiting extensive migratory behaviors in quest of sustenance and suitable breeding grounds. Demersal species, such as cod and halibut, inhabit the ocean floor, where they forage on benthic organisms amidst rocky reefs or sandy substrates. Meanwhile, benthic species like anglerfish and flatfish have evolved mechanisms for camouflaging themselves and ambushing prey on the seabed.

Coastal habitats, encompassing estuaries, salt marshes, mangroves, and rocky shores, emerge as cradles of biodiversity and productivity within the aquatic realm. Serving as vital nurseries, feeding grounds, and breeding sites, these habitats support intricate food webs and provide indispensable ecosystem services. Estuaries, in particular, act as transitional zones between freshwater and marine realms, nurturing species like salmon and striped bass amidst nutrient-rich waters. Salt marshes and mangroves offer sanctuary to juvenile fish, shielding them from predators while providing ample sustenance. Rocky shores, characterized by a mosaic of microhabitats and tidal pools, host diverse fish species adapted to the rigors of this dynamic environment.

Freshwater habitats, inclusive of rivers, streams, lakes, and wetlands, host a plethora of fish species uniquely adapted to freshwater conditions. Rivers and streams, with their flowing waters and diverse substrates, accommodate species such as trout and salmon, renowned for their migratory behaviors and spawning rituals in natal streams. Lakes, characterized by varying depths and still or slow-moving waters, harbor species like bass and perch, occupying different zones within these aquatic ecosystems. Wetlands, comprising swamps, marshes, and bogs, emerge as hubs of biodiversity, nurturing various fish species, especially those adapted to shallow, vegetated waters.

The distribution of fish species is intricately influenced by a myriad of factors, encompassing both biotic and abiotic elements. Temperature emerges as a pivotal determinant of geographical range, shaping the distribution of species adapted to either cold-water or warm-water environments. Similarly, salinity plays a crucial role, with marine, freshwater, and estuarine species occupying distinct niches. Oxygen availability, food resources, habitat structure, competition, predation, and reproductive requisites further sculpt the distribution patterns of fish species, dictating their presence in specific habitats and influencing community dynamics.

Human activities, ranging from habitat destruction and pollution to overfishing and climate change, exert profound impacts on fish distribution and aquatic ecosystem health. Habitat degradation fragments habitats, disrupts migration routes, and diminishes fish populations, while pollution compromises water quality and habitat suitability. Overfishing destabilizes food webs, leading to cascading effects on ecosystem structure and function. Climate change alters temperature regimes, precipitation patterns, and oceanic conditions, reshaping the distribution of fish species and challenging their adaptability to changing environments.

Adaptations to freshwater environments

Adaptations to freshwater environments encompass a wide spectrum of physiological, anatomical, and behavioral traits enabling organisms to flourish in habitats like rivers, lakes, and streams. Such ecosystems pose distinct challenges, including variations in water flow, temperature, pH, and nutrient availability. Organisms, to thrive and propagate in these fluid environments, have developed specialized adaptations facilitating efficient resource utilization, predator avoidance, and coping with environmental oscillations. Osmoregulation stands as a pivotal challenge in freshwater environments, necessitating the regulation of water and ion balance within organisms' bodies. In contrast to marine settings where dehydration is a constant threat due to seawater's higher osmotic pressure, freshwater environments present the inverse challenge of excessive water uptake. Thus, organisms in freshwater habitats must actively intake salts while restricting water absorption. This feat is achieved through specialized ion transport mechanisms, notably in the gills and kidneys. Freshwater fish exemplify this through active ion uptake, preventing osmotic imbalance-induced cellular and tissue swelling.

Temperature regulation is another critical facet of freshwater adaptation. Unlike the relatively stable temperatures of marine settings, freshwater habitats undergo significant temperature fluctuations driven by seasonal changes, sunlight exposure, and terrestrial influences. Organisms in these ecosystems must either tolerate a broad temperature range or possess mechanisms to seek out suitable thermal conditions. Many freshwater fish, being ectothermic, adjust their body temperature with the environment. Additionally, physiological adaptations like antifreeze proteins aid survival in icy waters, while others tolerate warmer tropical conditions. Nutrient availability also profoundly shapes freshwater adaptation. While these ecosystems can be nutrient-rich, the spatial and temporal distribution of these resources varies. Thus, organisms must efficiently exploit these resources for growth, reproduction, and energy. Various feeding strategies have evolved for this purpose, including filter feeding, detritivory, herbivory, and predation, each maximizing nutrient intake in unique ways.

Respiration in freshwater environments poses challenges due to lower oxygen concentrations compared to air and marine habitats. Organisms have developed specialized respiratory adaptations to cope with oxygen level fluctuations, such as highly vascularized gills and accessory respiratory structures like labyrinth organs in fish, and specialized respiratory structures in aquatic invertebrates. Physical adaptations in freshwater environments involve modifications to body shape, size, and coloration enhancing survival and ecological success. Streamlined body shapes reduce drag, aiding swimming efficiency, while countershading provides camouflage against predators. Mimicry is also observed, enhancing protection or hunting success.

Behavioral adaptations are crucial for resource exploitation, predator avoidance, and effective navigation. Migration, habitat selection, and complex reproductive behaviors are common. For instance, migration patterns like anadromy and catadromy are prevalent among fish species, and complex reproductive behaviors include courtship displays, nest-building, and parental care.

1.5 MARINE FISHES

Common marine fish species

Common marine fish species represent a diverse array of aquatic organisms found in oceans and seas globally, playing pivotal roles in marine ecosystems and supporting coastal economies. Understanding their characteristics, habitats, and ecological roles is vital for comprehending the complexity and biodiversity of ocean environments. Among the most iconic and diverse groups of marine fish are coral reef fish, residing in tropical waters' vibrant and biodiverse ecosystems. Renowned for their stunning colors and unique behaviors, these fish include butterflyfish, often found in pairs or small groups feeding primarily on coral polyps. Another group, angelfish, contributes to reef health by controlling algae growth. Clownfish, known for their symbiotic relationship with sea anemones, provide protection and aid in attracting prey, showcasing intricate interdependencies within coral reef ecosystems.

Reef-associated species, thriving in complex coral reef environments, play vital ecological roles as herbivores, carnivores, and omnivores. Parrotfish and surgeonfish help control algal growth, while predators like groupers and snappers maintain reef community balance. Additionally, these fish support valuable commercial and recreational fisheries, but overfishing and destructive practices threaten their populations, necessitating sustainable management strategies. Moving into the open ocean, pelagic fish species exhibit migratory behaviors and form large schools. Iconic species like tuna, apex predators, are vital in pelagic food webs, while billfish, including marlin and swordfish, are sought after by recreational anglers. However, overfishing and bycatch pose threats to their populations. Other pelagic species like mackerel, mahi-mahi, and barracuda also contribute to pelagic ecosystems, serving as prey for larger predators. Apart from coral reefs and the open ocean, marine fish inhabit various coastal and offshore habitats such as rocky shores, sandy beaches, estuaries, and deep-sea trenches. Each habitat supports a unique assemblage of species adapted to specific environmental conditions. Rocky shores provide refuge for species like rockfish and wrasse, while sandy beaches host flatfish such as flounder. Estuaries support diverse species like mullet and striped bass, serving as essential feeding and nursery grounds. Deep-sea trenches and abyssal plains harbor unique and poorly understood fish communities.

Habitat and distribution

The interplay between habitat and distribution profoundly influences the ecology, behavior, and evolutionary trajectories of fish species worldwide. Fish exhibit remarkable adaptability, inhabiting a diverse array of aquatic environments, including freshwater rivers, lakes, marine ecosystems such as oceans, coral reefs, and estuaries. Comprehending the nuances of fish habitat preferences and distribution patterns is imperative for effective conservation, management, and sustainable utilization of these invaluable resources. Freshwater ecosystems encompass a mosaic of habitats, ranging from swiftly flowing rivers to tranquil lakes and intricate wetlands. These environments, characterized by lower salinity levels, offer rich food resources and conducive breeding grounds for various fish species. Rivers and streams, with their oxygen-rich, swiftly flowing waters, host a plethora of fish species adapted to different flow regimes and substrate types. Conversely, lakes and ponds provide still or slow-moving waters, with fish communities structured by factors such as depth, temperature, nutrient availability, and vegetation cover. Wetlands, acting as crucial transition zones between aquatic and terrestrial ecosystems, serve as essential nurseries and refuges for fish during different life stages. The distribution of freshwater fish is influenced by a multitude of factors, including habitat availability, water quality, temperature, flow regimes, and human activities such as habitat alteration and pollution. While some species exhibit wide distributions and habitat plasticity, others have more restricted ranges and specific habitat requirements. For instance, salmonids like trout and salmon gravitate towards cold, well-oxygenated streams with gravel bottoms for spawning, while species like catfish and carp thrive in warmer, turbid waters abundant with vegetation and organic matter.

Marine habitats, covering approximately 70% of the Earth's surface, harbor a staggering diversity of fish species adapted to various oceanic zones. Pelagic fish, such as tuna and mackerel, roam the open ocean, forming large schools to optimize foraging and predator avoidance. Benthic species like flounder and cod inhabit the seafloor, where they feed on invertebrates and detritus. Coral reefs, among the most biodiverse marine ecosystems, provide vital shelter, feeding grounds, and breeding sites for numerous fish species. Intertidal habitats, located between high and low tide marks, present unique challenges for fish, including exposure to air, temperature extremes, and predation. Intertidal fish species have evolved specialized behaviors and physiological adaptations to thrive in this dynamic environment, utilizing crevices, burrows, and tidal pools for shelter and refuge. Estuaries, where freshwater rivers meet the ocean, serve as critical nursery grounds and feeding areas for many fish species. These transitional zones, characterized by fluctuating salinity levels and nutrient-rich waters, offer shelter and protection for juvenile fish as they transition from freshwater to marine environments.

The distribution patterns of fish are influenced by a myriad of biogeographic factors, including historical events, geographic barriers, and evolutionary processes. Human activities such as habitat destruction, pollution, overfishing, and climate change pose significant threats to fish habitats and distribution patterns worldwide, underscoring the urgent need for concerted conservation efforts to safeguard these invaluable ecosystems and the myriad species they support.

Adaptations to marine environments

Adaptations to marine environments encompass a myriad of specialized traits and behaviors facilitating thriving amidst the ocean's challenges. From coastal shallows to abyssal depths, marine organisms have evolved remarkable coping mechanisms for salinity, pressure, temperature, and nutrient dynamics, illuminating the ocean's astonishing biodiversity and resilience. Osmoregulation stands as a cornerstone adaptation, ensuring salt and water balance within organisms. Marine life contends with maintaining osmotic equilibrium against the hypertonicity of seawater. Mechanisms like specialized gill cells in fish facilitate active ion transport, preventing dehydration while conserving water. Sea turtles, for instance, possess salt-excreting glands near their eyes, enabling them to ingest seawater without dehydration.

Buoyancy control is pivotal, especially for species occupying various water depths. Strategies like swim bladders in fish allow for precise buoyancy adjustments, aiding effortless ascent or descent in the water column. Cephalopods, such as squids, employ gas regulation or jet propulsion for buoyancy management. Temperature regulation is vital across diverse oceanic zones. Marine mammals like whales rely on insulating blubber to retain body heat in cold waters, while intertidal invertebrates like mussels employ shell closure or mucus secretion to endure temperature fluctuations during low tide. Pressure adaptation is imperative for deep-sea inhabitants facing extreme pressures. Specialized adaptations, including osmolytes and structural modifications, stabilize cells under pressure. Deep-sea organisms like anglerfish mitigate pressure changes with gelatinous bodies and reduced skeletal structures.

Behavioral adaptations are diverse, aiding survival and reproduction. Long-distance migrations, guided by complex cues like magnetic fields and chemical gradients, facilitate finding food or breeding grounds. Salmon exemplify this, navigating vast ocean currents to spawn in natal rivers using olfactory cues. Reproductive strategies vary, reflecting ecological niches and life histories. Pelagic fish release millions of eggs to offset high mortality rates, while some sharks give birth to live young for added protection. Asexual reproduction in marine invertebrates like corals ensures rapid habitat colonization. Symbiotic relationships are prevalent, enhancing survival through mutual benefits. Coral reefs exemplify this, where corals and photosynthetic algae foster symbiosis for nutrient exchange. Cleaner fish and shrimp provide services to larger fish in exchange for protection and food scraps.

Camouflage and mimicry aid evasion and predation. Cryptic coloration and behaviors help organisms blend with surroundings, such as the leafy seadragon resembling seaweed. Certain cephalopods can rapidly change skin color and texture to mimic their environment, evading predators. Efficient feeding strategies are vital in resource-limited oceans. Predators like sharks possess sensory adaptations for prey detection, while filter feeders like baleen whales consume plankton with specialized structures. Resilience to environmental change is paramount amid anthropogenic threats. Some corals exhibit tolerance to temperature and acidity changes, while others can regenerate damaged reefs. Yet, the rapid pace of human-induced changes underscores the urgent need for conservation and sustainable management efforts.

1.6 SUMMARY

The taxonomic categorization of fish stands as a comprehensive endeavor, encompassing a rich tapestry of species intricately adapted to various aquatic habitats across the globe. This classification system draws from multiple facets, including anatomical structures, genetic affinities, ecological roles, and behavioral patterns. Among the foundational divisions, three principal groups emerge: Agnatha (jawless fish), Chondrichthyes (cartilaginous fish), and Osteichthyes (bony fish), each delineating significant evolutionary trajectories within piscine lineage while manifesting distinctive attributes.

Agnatha, comprising jawless fish, represents a primitive cohort within the vertebrate realm, distinguished by the absence of true jaws and characterized by circular mouths adorned with rasping teeth. This group bifurcates into two primary taxa: lampreys, notorious for their parasitic feeding habits as they latch onto hosts to siphon blood and tissues, and hagfish, serving as scavengers vital to the recycling of marine detritus. Despite their antiquated visage, jawless fish wield considerable ecological significance, offering pivotal insights into vertebrate evolutionary narratives.

Chondrichthyes, typified by cartilaginous skeletons, encompass sharks, rays, and skates, among others. Endowed with well-honed senses, including acute vision, a discerning lateral line system, and electroreceptors facilitating the detection of electrical fields, cartilaginous fish often occupy apex predator roles, playing indispensable roles in marine ecosystem equilibrium. Sharks, with their formidable jaws and serrated dentition, emerge as preeminent hunters, while rays and skates, boasting flattened physiques conducive to benthic existence, prey upon bottom-dwelling organisms like crustaceans and mollusks.

Osteichthyes, the most diverse assemblage of fish, encompassing over 95% of all species, feature bony skeletons and possess swim bladders for buoyancy regulation. This group subdivides into Actinopterygii (ray-finned fish) and Sarcopterygii (lobe-finned fish). Ray-finned fish, exemplified by salmon, trout, and goldfish, exhibit fins supported by bony rays and constitute the majority of extant fish taxa. In contrast, lobe-finned fish such as coelacanths and lungfish, characterized by fleshy, lobed fins reminiscent of terrestrial vertebrate limbs, represent a distinct evolutionary lineage.

Within these principal groups, further taxonomic delineations occur, elucidating shared anatomical traits and evolutionary affinities. Hierarchical classifications ranging from orders to species furnish a framework for comprehending the extensive piscine diversity and evolutionary trajectories. Additionally, habitat-based classifications discern fish as freshwater, marine, or brackish, contingent upon their salinity preferences. Feeding habits further stratify fish into herbivorous, carnivorous, or omnivorous categories, reflecting dietary predilections and ecological roles.

Reproductive strategies also influence fish classification, with oviparous species laying externally fertilized eggs and viviparous species birthing live young nurtured through internal gestation. Moreover, ecological roles within ecosystems categorize fish as pelagic or benthic based on habitat preferences within the water column. Pelagic species inhabit open water and often form schools, while benthic species reside near or on the water bottom, engaging in bottom-dwelling prey consumption. Advancements in genetic analysis and molecular techniques continually refine fish taxonomy, unveiling intricate evolutionary relationships and resolving taxonomic ambiguities. Phylogenetic studies leveraging genetic data illuminate unforeseen evolutionary connections, augmenting our understanding of piscine diversity and evolutionary dynamics.

CHAPTER 2

FISHERIES IN INDIA

2.1 OVERVIEW OF FISHERIES IN INDIA

The historical trajectory of fisheries in India reflects a deep-rooted reliance on aquatic resources spanning millennia, intertwined with the nation's maritime legacy. India's expansive coastline, intricate river networks, and diverse aquatic ecosystems have historically sustained a robust fishing industry, evidenced by archaeological findings dating back to ancient civilizations like the Indus Valley Civilization and the Harappan culture. Over time, fishing techniques have evolved, with coastal and inland communities honing specialized skills for fishery activities. During the colonial era, the introduction of modern fishing technologies by European powers catalyzed significant transformations in India's fisheries. British colonial administrators recognized the economic potential of the country's marine resources, leading to the establishment of commercial fishing ventures. This period saw the advent of large-scale mechanized fishing, marked by the introduction of trawlers and purse seiners, albeit at the expense of displacing and marginalizing traditional fishing communities.

Post-independence, India witnessed further evolution in its fisheries sector, driven by governmental emphasis on economic development and food security. Institutions like the Central Marine Fisheries Research Institute (CMFRI) and the Department of Fisheries were established to modernize and regulate the industry. Concurrently, initiatives such as fishery cooperatives and government schemes aimed to uplift fishing communities' socio-economically while enhancing sector productivity and efficiency. The Green Revolution of the 1960s and 1970s, while boosting agricultural output, inadvertently impacted fisheries and aquatic ecosystems. Wetlands and coastal habitats were converted for agricultural use, leading to habitat loss and dwindling fish populations. Pollution from agricultural runoff and industrial effluents compounded these challenges, jeopardizing water quality and aquatic biodiversity.

Despite these obstacles, India's fisheries sector has thrived, contributing significantly to the economy, food security, and employment. Coastal states like Kerala, Tamil Nadu, Andhra Pradesh, Gujarat, and Maharashtra emerge as key fish producers, boasting vibrant fishing communities and flourishing seafood industries. Marine fisheries dominate production, encompassing species like Indian mackerel, sardines, shrimp, and tuna, while inland fisheries, including freshwater aquaculture and capture fisheries, also play pivotal roles. Aquaculture emerges as a vital facet of India's fisheries, offering avenues for diversification, intensification, and value addition. Practices like pond culture, cage culture, and integrated multi-trophic aquaculture (IMTA) have proliferated, buoyed by rising demand for fish protein and favorable governmental policies. Carp species dominate freshwater aquaculture, while shrimp farming, notably Indian white shrimp (Penaeus indicus) and Pacific white shrimp (*Litopenaeus vannamei*), reigns along the coast.

However, the fisheries sector grapples with multifaceted challenges jeopardizing its sustainability. Overexploitation of fish stocks, habitat degradation, illegal fishing, and resource conflicts necessitate urgent attention. Climate change exacerbates these issues, posing threats like rising sea levels and extreme weather events. Inadequate governance, enforcement, and infrastructure further impede efforts toward sustainable management and inclusive development. To address these challenges, collaborative efforts involving government, civil society, research institutions, and fishing communities are imperative. The National

Fisheries Policy (NFP) of 2020 underscores commitments to responsible fisheries management, biodiversity conservation, and equitable socio-economic development. Embracing ecosystem-based approaches, marine spatial planning, and sustainable aquaculture practices are pivotal for safeguarding marine ecosystems and fostering community resilience.

The COVID-19 pandemic has posed unprecedented challenges, disrupting fisheries' supply chains, markets, and labor. Government interventions aim to alleviate socio-economic impacts and bolster sectoral recovery and resilience. Moving forward, prioritizing sustainable fisheries management, climate adaptation, and inclusive development remains paramount. Strengthening governance, stakeholder engagement, and partnerships are crucial for realizing goals of food security, poverty alleviation, and environmental sustainability. Empowering marginalized fishing communities, particularly women, is essential for enhancing sectoral resilience in the face of emerging challenges."

2.2 MAJOR FISH SPECIES IN INDIA

In India, fish species wield considerable influence on the nation's economy, cultural practices, and dietary habits, catering to the sustenance of millions. Among these piscine entities, the carp species stand out prominently, encompassing Rohu (*Labeo rohita*), Catla (*Catla catla*), Mrigal (*Cirrhinus mrigala*), and others. These species hold immense value owing to their nutritive profiles, rapid growth kinetics, and versatility in adapting to diverse aquatic milieus. Carps proliferate in ponds, reservoirs, and rivers throughout India, thereby significantly bolstering both freshwater aquaculture and wild fisheries.

***Labeo rohita*,** commonly known as Rohu, assumes a paramount role among Indian carp species. Indigenous to the Indian subcontinent, this freshwater fish inhabits rivers, reservoirs, and ponds ubiquitously. Renowned for its delectable flesh, Rohu enjoys widespread consumption across the nation. Its commendable growth rates and adeptness in diverse environmental settings render it a favored choice for aquaculture endeavors. Rohu, predominantly a benthic feeder, subsists on detritus, algae, and small invertebrates, often co-cultured with other carp species like Catla and Mrigal.

Catla catla, or simply Catla, represents another significant carp species in India. Native to the Indian subcontinent, Catla thrives in rivers, lakes, and reservoirs across the nation. Celebrated for its rapid growth and succulent flesh, Catla attains substantial sizes within short intervals. Preferring surface feeding, Catla predominantly consumes phytoplankton, aquatic plants, and zooplankton. Its inclusion in composite fish culture systems enhances the overall productivity and economic viability of aquaculture operations alongside other carp species.

Cirrhinus mrigala, recognized as Mrigal, constitutes a widely cultivated carp species in India. Indigenous to freshwater bodies like rivers, lakes, and ponds within the Indian subcontinent, Mrigal is esteemed for its swift growth, environmental adaptability, and superior flesh quality. As an omnivorous feeder, Mrigal partakes in a varied diet comprising plant and animal matter, including detritus, plankton, and aquatic vegetation. Often raised in polyculture systems, Mrigal contributes significantly to integrated fish farming practices, complementing other carp species.

Beyond carps, India harbors several other noteworthy fish species pivotal to its fisheries and aquaculture sectors. One such exemplar is *Tenualosa ilisha,* colloquially known as Hilsa or ilish, distinguished for its sumptuous flavor and tender flesh. Predominantly inhabiting the rivers and estuaries along the eastern coast, particularly within the Ganges-Brahmaputra-Meghna river system, Hilsa undertakes anadromous migrations from sea to freshwater rivers for spawning purposes. A culinary favorite, Hilsa finds its way into traditional dishes such as Hilsa curry and

Hilsa fry, delighting seafood enthusiasts.

Mahseer stands as another iconic fish species endemic to India, acclaimed for its formidable size, vigor, and sporting allure. Belonging to the genus Tor, Mahseer comprises several species distributed across the rivers and streams of the Indian subcontinent. Notably, Tor putitora, or the Golden Mahseer, enthralls anglers with its imposing dimensions and tenacious resistance. Revered in folklore and traditional angling customs, Mahseer faces threats from overfishing and habitat degradation, necessitating conservation endeavors to safeguard these emblematic fish species. Indian major carp species like Rohu, Catla, and Mrigal, alongside significant fish species such as Hilsa and Mahseer, assume pivotal roles in India's fisheries and aquaculture sectors. Serving as indispensable sources of protein and essential nutrients, they underpin the sustenance of millions while fostering livelihoods for fishermen and fish farmers. Furthermore, they contribute substantially to the nation's cultural heritage. Hence, sustainable management and conservation initiatives are imperative to ensure the perpetual abundance and accessibility of these invaluable fish species for future generations.

2.3 FISHERIES PRODUCTION BY STATE

India's fisheries sector presents a captivating mosaic, intricately woven from diverse marine and freshwater ecosystems. From the sun-drenched coastlines to the life-giving rivers and vibrant wetlands, fisheries play a pivotal role in the nation's socio-economic well-being, ensuring food security, providing essential nutrients, and generating significant economic activity and livelihood opportunities. Analyzing fisheries production across individual states reveals the regional variations and strengths that contribute to India's position as a major player in global fisheries.

**Dominant Southern Trio: Andhra Pradesh, Tamil Nadu, and

Kerala: Undeniably, the southern states of Andhra Pradesh, Tamil Nadu, and Kerala emerge as the powerhouses of India's fisheries sector. Andhra Pradesh reigns supreme, consistently topping national charts in fish production due to a confluence of favorable factors. Its extensive coastline exceeding 970 kilometers serves as a nursery and feeding ground for a rich diversity of marine resources, including commercially valuable species like prawns, shrimps, pomfrets, and hilsa. Additionally, Andhra Pradesh has prioritized aquaculture development, fostering a flourishing brackishwater aquaculture sector and a continuously expanding network of inland fish ponds. Tamil Nadu, another coastal powerhouse, leverages its 900 kilometers of coastline to its advantage, boasting abundant catches of tuna, mackerel, and seerfish. Aquaculture thrives in the fertile Cauvery delta region, contributing significantly to the state's overall fisheries production. Kerala, renowned for its idyllic backwaters and serene beaches, exhibits leadership in both marine and inland fisheries. Marine resources like sardines, prawns, and ribbonfish contribute substantially, while traditional practices like 'Chinesevala Karunya' (cage culture) and 'Pokkali' (prawn farming integrated with paddy fields) are hallmarks of Kerala's inland fisheries, showcasing a unique blend of tradition and innovation.

Western Littoral Powerhouses: Gujarat, Maharashtra, Goa, and Karnataka: Moving westward, the states of Gujarat, Maharashtra, Goa, and Karnataka paint a vibrant picture of marine fisheries production. Gujarat, with its impressive 1600 kilometers of coastline, is a major producer of commercially valuable fish like pomfrets, catfish, and hilsa. The state has also made significant strides in cage aquaculture, particularly in saline waters, demonstrating a commitment to exploring sustainable and productive fishing practices. Maharashtra, boasting a long coastline stretching over 720 kilometers, contributes a rich diversity of marine fish to the national catch, including Bombay duck, prawns, and sharks. Inland fisheries also contribute significantly, with reservoirs, lakes, and diverse aquaculture practices forming the backbone of the state's freshwater fisheries sector. Goa, known for its pristine beaches and delectable seafood, witnesses a seasonal surge in fish landings during the monsoon months. Kingfish, mackerel, and sardines are major catches, while aquaculture in ponds and cages is gaining traction, potentially diversifying Goa's fisheries production in the future. Karnataka, with its 300 kilometers of coastline, contributes to the national fish basket with catches of mackerel, sardines, and prawns. Inland fisheries, particularly in the Cauvery basin, are crucial for the state's fish production, highlighting the importance of freshwater ecosystems for sustaining India's fisheries sector.

Eastern Shores: West Bengal, Odisha, and Andhra Pradesh: The eastern coastline of India, encompassing West Bengal, Odisha, and Andhra Pradesh (partially), presents a unique picture of fisheries production. West Bengal, despite having a relatively shorter coastline of around 158 kilometers, emerges as a major producer of fish owing to its extensive network of rivers and estuaries, particularly the mighty Ganges and Brahmaputra. Hilsa, a prized catch with significant economic and cultural value, migrates up these rivers, forming the backbone of West Bengal's inland fisheries. Additionally, brackishwater aquaculture, particularly for prawns and shrimp, is flourishing in the state, demonstrating the potential for integrating marine and freshwater resources for sustainable fisheries development. Odisha, with a coastline exceeding 480 kilometers, witnesses significant marine fish landings, especially of commercially valuable species like mackerel, prawns, and pomfrets. Chilika Lake, Asia's largest brackishwater lagoon, plays a vital role in Odisha's fisheries sector. This unique ecosystem supports a thriving environment for fish and prawn culture, highlighting the importance of protecting and managing wetlands for sustainable fisheries production.

Landlocked Leaders: Punjab, Haryana, and Uttar Pradesh: While often overshadowed by coastal states, landlocked states like Punjab, Haryana, and Uttar Pradesh also contribute significantly to India's overall fisheries production. Punjab, with its network of canals and reservoirs, has witnessed a boom in aquaculture, particularly focusing on carp culture. This shift towards aquaculture practices demonstrates the ability of landlocked states to adapt and contribute to the national fisheries sector despite limitations in natural water resources. Similarly, Haryana, despite limited water resources, has made strides in developing aquaculture, focusing on carps and catfish in ponds and tanks, showcasing potential for further development in this domain.

2.4 FRESHWATER FISHERIES IN INDIA

India stands as a beacon of freshwater abundance, fostering a thriving fishery sector essential for economic sustenance and food security. Diverse aquatic habitats, from rivers to reservoirs, ponds, and inland water bodies, provide the stage for this flourishing industry. Emphasizing sustainable growth, freshwater aquaculture has emerged as a linchpin, promising enhanced production without compromising ecological integrity. Notably, key regions such as the Gangetic plains, Brahmaputra Valley, Cauvery Delta, and Krishna-Godavari Basin epitomize India's aquatic wealth, each boasting unique hydrological nuances conducive to diverse fish cultivation methodologies. The Gangetic plains, spanning Uttar Pradesh, Bihar, and West Bengal, stand as a bastion of freshwater fisheries. Its intricate network of rivers and floodplains provides an ideal backdrop for cultivating prized species like rohu, catla, and mrigal. Employing traditional earthen ponds, farmers implement semi-intensive polyculture practices, harnessing organic and inorganic fertilizers to augment natural food production. Modern interventions, including improved fish strains and management techniques, further amplify productivity, ensuring a sustainable harvest.

Venturing eastward, the Brahmaputra Valley in Assam emerges as another pivotal fisheries hub. The labyrinthine river and expansive floodplain wetlands nurture a diverse array of species, including common carp and grass carp alongside indigenous favorites. Pond-based aquaculture, often integrated with agriculture, dominates the landscape, fostering synergies between fish culture and farming activities. Vigilant water quality monitoring and advanced hatchery operations underscore the region's commitment to sustainable growth. Meanwhile, the Cauvery Delta in Tamil Nadu showcases advanced aquaculture paradigms, blending traditional wisdom with modern techniques. Here, Indian major carps coexist with tilapia and catfish, thriving in meticulously managed earthen and lined ponds. Aeration systems, formulated feeds, and biofloc technology exemplify the region's dedication to optimizing production while minimizing environmental impact. The Krishna-Godavari Basin, spanning Andhra Pradesh and Telangana, epitomizes innovation in freshwater fisheries. Boasting a diverse aquatic landscape, from rivers to reservoirs, the basin hosts a myriad of species including pangasius and tilapia. From extensive traditional systems to highly intensive modern setups, farmers leverage advanced technologies to bolster productivity. Reservoir fisheries further bolster fish production, supporting both commercial enterprises and local communities.

In tandem with these regional endeavors, governmental initiatives and research institutions like the Indian Council of Agricultural Research (ICAR) and the Central Institute of Freshwater Aquaculture (CIFA) spearhead sustainable aquaculture practices. Research spanning breeding, nutrition, and disease management informs farmers, while training programs and extension services disseminate best practices nationwide. Diverse methodologies underpin India's freshwater aquaculture, catering to varied environmental contexts and resource availabilities. Pond culture, the cornerstone of Indian aquaculture, facilitates controlled rearing environments, while polyculture maximizes resource efficiency. Cage culture harnesses natural water bodies, particularly suitable for high-value species, while integrated farming fosters synergies between aquaculture and agriculture, enhancing productivity and sustainability.

For cutting-edge efficiency, recirculating aquaculture systems (RAS) offer a closed-loop solution, minimizing water usage and environmental discharge. Despite higher initial costs, RAS promises unparalleled sustainability and production efficiency, particularly for high-value species. In essence, India's freshwater fisheries epitomize a delicate balance between tradition and innovation, sustenance and growth. As the nation charts a path towards sustainable development, prioritizing innovation, environmental stewardship, and community engagement will be imperative for ensuring the longevity and prosperity of this critical industry.

2.5 MARINE FISHERIES IN INDIA

Marine fisheries constitute a pivotal sector within India, exerting significant influence on the nation's economy, food security, and employment landscape, thereby underpinning the livelihoods of millions. The expansive coastline spanning over 7,500 kilometers, coupled with an exclusive economic zone (EEZ) extending to 2.02 million square kilometers, fosters a diverse array of marine resources. The marine fisheries sector in India is distinguished by its multifaceted biodiversity, characterized by a myriad of species harvested through a blend of traditional and contemporary methodologies, which have evolved under the influence of natural phenomena, technological progressions, and socio-economic dynamics. India's marine fishing domain is segmented into three principal regions: the west coast, the east coast, and the Andaman and Nicobar Islands. The west coast, encompassing states such as Gujarat, Maharashtra, Goa, Karnataka, and Kerala, boasts highly productive fishing grounds due to the synergy of the southwest monsoon and nutrient-rich upwelling waters. Prominent fishing zones along this coast include the Gulf of Kutch, Gulf of Khambhat, Konkan coast, and Malabar coast. In contrast, the east coast, comprising Tamil Nadu, Andhra Pradesh, Odisha, and West Bengal, benefits from the northeast monsoon and nutrient influx from major river systems, housing significant fishing areas like the Coromandel coast, Andhra coast, and Orissa coast. The Andaman and Nicobar Islands offer rich fishing grounds characterized by unique marine ecosystems, relatively less exploited compared to mainland regions.

The spectrum of species harvested in Indian marine fisheries is diverse, spanning various fish, crustaceans, mollusks, and other marine organisms. Economically pivotal species include Indian mackerel, oil sardine, threadfin breams, ribbonfish, croakers, tuna, and seerfish. Crustaceans such as shrimp and crabs constitute a substantial portion of the marine catch, alongside mollusks like cuttlefish, squid, and octopus, valued for both domestic consumption and export. Additionally, high-value species like groupers, pomfrets, and hilsa shad cater to discerning market demands and culinary preferences. The methodologies employed in Indian marine fisheries range from traditional artisanal techniques to modern, mechanized approaches. Traditional methods involve the utilization of small boats such as catamarans and plank-built vessels, equipped with gear like gill nets, cast nets, hand lines, and traps, which typically have minimal environmental impact. Conversely, modern techniques involve mechanized vessels outfitted with advanced gear, including trawlers, purse seiners, and longliners, targeting demersal and pelagic species through methods such as trawling, purse seining, and longlining.

The development and sustenance of marine fisheries in India encounter multifaceted challenges, including overfishing, habitat degradation, pollution, and climate change impacts. Regulatory interventions such as fishing quotas, seasonal closures, and marine protected areas (MPAs) are pivotal in addressing overfishing and fostering sustainability. Concurrently, efforts to mitigate habitat degradation, combat pollution, and adapt to climate change are imperative for safeguarding marine ecosystems and securing the continuity of fisheries. The cultural and social dimensions of marine fisheries in India are deeply entrenched within fishing communities, characterized by rich traditions and ecological knowledge. The integration of traditional ecological knowledge (TEK) with scientific research and modern management practices enhances the efficacy of fisheries management, underscoring the significance of community engagement and participatory approaches.

Endeavors to enhance the sustainability and resilience of Indian marine fisheries encompass community-based management, capacity building, and technological innovations. Community-based management fosters stakeholder engagement in decision-making processes, while capacity-building initiatives equip fishers with skills for sustainable practices and alternative livelihoods. Technological innovations, such as selective fishing gear and fish aggregating devices, contribute to sustainability by reducing bycatch and enhancing habitat preservation. International cooperation through regional fisheries management organizations (RFMOs) and adherence to international agreements play crucial roles in sustainable fisheries management and conservation efforts. Research, monitoring, education, and integrated planning initiatives further bolster the sustainability and resilience of marine fisheries, ensuring their enduring contribution to India's economy, culture, and food security.

2.6 SUMMARY

The fisheries sector in India stands as a dynamic and vital contributor to the nation's economy, food security, and employment landscape, reflecting its rich abundance of fishery resources bestowed by extensive coastlines, rivers, lakes, and reservoirs. This article presents an overview of the status and distribution of fisheries in India, focusing on historical development, regional diversity, species composition, economic significance, challenges, and policy frameworks governing the sector. Historically, fishing has been deeply ingrained in Indian culture and livelihoods, evolving from traditional practices to a significant economic force through technological advancements, increased investment, and government support. Mechanized fishing boats, modern gear, and improved post-harvest practices have boosted fish production and productivity significantly.

India's marine fisheries are distributed along its vast coastline, divided into western and eastern coasts, each harboring distinct ecological characteristics and fishery resources. The western coast, comprising states like Gujarat, Maharashtra, Karnataka, Goa, and Kerala, boasts rich marine biodiversity and abundant fishing grounds. Gujarat and Kerala particularly stand out as major contributors to marine fish production, owing to well-developed infrastructure and large fishing fleets. On the eastern coast, states such as Tamil Nadu, Andhra Pradesh, Odisha, and West Bengal play pivotal roles in the regional economy, with extensive estuarine and deltaic systems supporting diverse fisheries. Tamil Nadu and Andhra Pradesh lead in marine fish production, investing significantly in modernizing fishing fleets and expanding processing facilities.

Inland fisheries, encompassing rivers, reservoirs, lakes, ponds, and wetlands, provide livelihoods to millions and contribute significantly to national fish production. Major river systems like the Ganges, Brahmaputra, and others, along with numerous smaller rivers, support diverse fish populations. States like Uttar Pradesh, Bihar, West Bengal, Andhra Pradesh, and Assam are prominent players in inland fisheries. Aquaculture, encompassing freshwater, brackish water, and marine practices, is a rapidly growing sector in Indian fisheries. Freshwater aquaculture dominates, with states like Andhra Pradesh, West Bengal, Tamil Nadu, and Odisha leading in production. Brackish water aquaculture, particularly shrimp farming, has gained prominence, with Andhra Pradesh being a major producer. Marine aquaculture shows potential for growth, especially in coastal states like Kerala, Tamil Nadu, and Maharashtra. The economic significance of the fisheries sector in India is profound, providing direct and indirect employment, ensuring food security, and contributing to foreign exchange earnings through exports. Despite its importance, the sector faces challenges such as overfishing, habitat degradation, climate change impacts, IUU fishing, and infrastructural limitations. Effective fisheries management, conservation efforts, climate adaptation strategies, and infrastructural development are crucial for sustainable growth. Government initiatives like the Blue Revolution and Pradhan Mantri Matsya Sampada Yojana aim to enhance fish production, improve livelihoods, and promote sustainable practices. Research institutions like the Indian Council of Agricultural Research (ICAR) and others play vital roles in advancing the sector through innovation and addressing emerging challenges.

CHAPTER 3

ECONOMICS OF FISHERIES IN HARYANA

3.1 FISHERIES INDUSTRY IN INDIA

The fisheries sector in India stands as a dynamic and vital contributor to the nation's economy, food security, and employment landscape, reflecting its rich abundance of fishery resources bestowed by extensive coastlines, rivers, lakes, and reservoirs. This article presents an overview of the status and distribution of fisheries in India, focusing on historical development, regional diversity, species composition, economic significance, challenges, and policy frameworks governing the sector. Historically, fishing has been deeply ingrained in Indian culture and livelihoods, evolving from traditional practices to a significant economic force through technological advancements, increased investment, and government support. Mechanized fishing boats, modern gear, and improved post-harvest practices have boosted fish production and productivity significantly.

India's marine fisheries are distributed along its vast coastline, divided into western and eastern coasts, each harboring distinct ecological characteristics and fishery resources. The western coast, comprising states like Gujarat, Maharashtra, Karnataka, Goa, and Kerala, boasts rich marine biodiversity and abundant fishing grounds. Gujarat and Kerala particularly stand out as major contributors to marine fish production, owing to well-developed infrastructure and large fishing fleets. On the eastern coast, states such as Tamil Nadu, Andhra Pradesh, Odisha, and West Bengal play pivotal roles in the regional economy, with extensive estuarine and deltaic systems supporting diverse fisheries. Tamil Nadu and Andhra Pradesh lead in marine fish production, investing significantly in modernizing fishing fleets and expanding processing facilities.

Inland fisheries, encompassing rivers, reservoirs, lakes, ponds, and wetlands, provide livelihoods to millions and contribute significantly to national fish production. Major river systems like the Ganges, Brahmaputra, and others, along with numerous smaller rivers, support diverse fish populations. States like Uttar Pradesh, Bihar, West Bengal, Andhra Pradesh, and Assam are prominent players in inland fisheries. Aquaculture, encompassing freshwater, brackish water, and marine practices, is a rapidly growing sector in Indian fisheries. Freshwater aquaculture dominates, with states like Andhra Pradesh, West Bengal, Tamil Nadu, and Odisha leading in production. Brackish water aquaculture, particularly shrimp farming, has gained prominence, with Andhra Pradesh being a major producer. Marine aquaculture shows potential for growth, especially in coastal states like Kerala, Tamil Nadu, and Maharashtra.

The economic significance of the fisheries sector in India is profound, providing direct and indirect employment, ensuring food security, and contributing to foreign exchange earnings through exports. Despite its importance, the sector faces challenges such as overfishing, habitat degradation, climate change impacts, IUU fishing, and infrastructural limitations. Effective fisheries management, conservation efforts, climate adaptation strategies, and infrastructural development are crucial for sustainable growth. Government initiatives like the Blue Revolution and Pradhan Mantri Matsya Sampada Yojana aim to enhance fish production, improve livelihoods, and promote sustainable practices. Research institutions like the Indian Council of Agricultural Research (ICAR) and others play vital roles in advancing the sector through innovation and addressing emerging challenges.

3.2 OVERVIEW OF FISHERIES IN HARYANA

The fisheries sector in Haryana has undergone a remarkable transformation, transitioning from traditional methods to modern aquaculture practices, reflecting the state's commitment to meet escalating fish demands and bolster its economy. Historically, fishing in Haryana traces back to ancient times, where communities relied on natural water bodies like rivers, lakes, and ponds for sustenance. As population pressures escalated and natural resources depleted, the focus shifted towards organized and sustainable fisheries management approaches. The advent of aquaculture marked a significant turning point in the historical trajectory of fisheries in Haryana. The establishment of fish farms and hatcheries ushered in a new era, with the state government recognizing the potential of aquaculture to boost fish production and rural income. Initiatives encompassed the construction of fish ponds, distribution of fingerlings, training programs for fish farmers, and the introduction of fish species suited to Haryana's climate.

Presently, the fisheries sector in Haryana thrives, buoyed by governmental interventions, technological advancements, and active engagement from fish farmers and entrepreneurs. Fish farming has emerged as a lucrative enterprise, contributing substantially to the state's economy and employment, particularly in rural areas. Haryana hosts a network of fish farms, hatcheries, and processing units, cultivating diverse fish species, both freshwater and brackish. A notable trend is the adoption of modern aquaculture practices to enhance productivity and sustainability. Fish farmers increasingly employ scientific techniques like pond management, water quality monitoring, and disease control measures. Innovations such as recirculating aquaculture systems (RAS), biofloc technology, and integrated fish farming are gaining traction for their efficiency and environmental benefits.

Governmental support and regulation play pivotal roles in the fisheries sector. Various departments and agencies oversee different facets of fisheries management, collaborating to promote sustainable aquaculture, conserve natural fish stocks, and offer support services to fish farmers. Challenges, including overfishing, habitat loss, pollution, and climate change, threaten the sector's long-term viability. In response, the government has implemented conservation and management measures, establishing protected areas, sanctuaries, and breeding grounds. Enforcement of fishing regulations aids in preventing overexploitation and promoting responsible fishing practices. Value addition and market development are key components of fisheries management. Interventions such as fish processing units, cold storage facilities, and marketing networks enhance product quality and marketability, offering opportunities for income diversification and access to higher-value markets for fish farmers.

3.3 DISTRICT WISE FISH PRODUCTION IN HARYANA

Fish production in Haryana is an important aspect of the state's economy and plays a significant role in meeting the protein needs of its population. While Haryana is primarily known for its agricultural output, the aquaculture sector has been gaining momentum in recent years, with several districts actively involved in fish production. Understanding the district-wise distribution of fish production, along with production statistics and comparisons, provides valuable insights into the state's aquaculture industry. Among the major districts involved in fish production in Haryana, Kurukshetra, Ambala, and Yamunanagar stand out as key contributors to the state's aquaculture sector. These districts benefit from favorable geographical conditions, availability of water resources, and supportive government initiatives aimed at promoting fish farming.

Kurukshetra, located in the northern part of Haryana, has emerged as a prominent hub for fish production, with a large number of fish farms and hatcheries operating in the region. The district's proximity to the Yamuna river and its tributaries provides abundant water for fish farming activities, making it an ideal location for aquaculture development.

Ambala district, situated in the southeastern part of Haryana, is another important player in the state's fish production landscape. With its extensive network of canals, reservoirs, and ponds, Ambala offers favorable conditions for freshwater fish farming. The district has seen a steady growth in fish production in recent years, driven by increasing demand for fish products and improved aquaculture practices. Yamunanagar, located in the western part of Haryana, is also actively involved in fish farming activities. The district's proximity to the Yamuna river and its well-developed irrigation infrastructure make it conducive to aquaculture development. Production statistics provide valuable insights into the scale and magnitude of fish production in different districts of Haryana. Kurukshetra, being a major center for fish farming, contributes significantly to the state's overall fish production. The district boasts a large number of fish farms and hatcheries, producing a diverse range of fish species including carp, catfish, and tilapia. Ambala district also plays a significant role in fish production, with a growing number of fish farmers adopting modern aquaculture techniques to enhance productivity. Yamunanagar, though relatively smaller in size compared to Kurukshetra and Ambala, is emerging as a promising destination for fish farming, with increasing investments in aquaculture infrastructure and technology.

Comparing fish production across different districts of Haryana provides insights into the relative contributions of each region to the state's aquaculture sector. While Kurukshetra leads in terms of overall fish production, Ambala and Yamunanagar are also making significant strides in this regard. Factors such as availability of water resources, land suitability, market demand, and government support influence the distribution of fish production across districts. Kurukshetra's strategic location and well-established aquaculture infrastructure give it a competitive edge in fish farming, attracting investments and fostering growth in the sector. Ambala and Yamunanagar, though smaller in scale, are leveraging their natural resources and geographic advantages to expand their fish farming activities and contribute to the state's food security and economic development. In addition to freshwater fish production, Haryana also has potential for the development of brackish water and marine aquaculture. Coastal districts like Ambala and Yamunanagar offer opportunities for shrimp farming and other marine species cultivation, leveraging their proximity to the Arabian Sea and the Bay of Bengal. Brackish water aquaculture, which involves farming species that thrive in partially saline conditions, can be developed in coastal areas and estuaries, further diversifying Haryana's aquaculture portfolio and boosting rural livelihoods.

1. **Ambala:** Notable for its rivers and water bodies, Ambala's fish production contributes significantly to the state's total output.

2. **Kurukshetra:** With its extensive network of canals, Kurukshetra has seen a steady increase in fish farming and production.

3. **Karnal:** The district's focus on integrated farming systems has boosted its fish production in recent years.

4. **Panipat:** Known for its reservoirs, Panipat's fish production is a vital part of its agricultural sector.

5. **Hisar:** Hisar's large ponds and tanks have made it a key player in Haryana's fishery industry.

6. **Rohtak:** The district's emphasis on sustainable aquaculture practices has resulted in a consistent fish yield.

7.   **Jind:** With initiatives to expand water bodies, Jind's fish production is on the rise.

8.   **Bhiwani:** Bhiwani's focus on enhancing fish breed quality has positively impacted its production levels.

9.   **Sirsa:** The district's innovative aquaculture techniques have led to an increase in fish production.

10. **Fatehabad:** Known for its freshwater fish, Fatehabad has seen growth in both production and variety.

11. **Mahendragarh:** The district's strategic location has facilitated the distribution and growth of its fishery sector.

12. **Rewari:** Rewari's commitment to fishery development programs has bolstered its production.

13. **Gurgaon:** With modern aquaculture facilities, Gurgaon's fish production is thriving.

14. **Faridabad:** Faridabad's diverse fish species contribute to the district's robust fishery economy.

15. **Mewat:** Mewat's focus on rural fishery development has led to a substantial increase in production.

16. **Palwal:** Palwal's investment in fishery infrastructure has paid off with higher production figures.

17. **Panchkula:** The district's natural resources support a healthy fish production.

18. **Yamunanagar:** Situated along the Yamuna river, the district has a rich tradition of fish farming.

19. **Sonipat:** Sonipat's strategic initiatives have made it a significant contributor to the state's fishery sector.

3.4 ECONOMIC IMPACT OF FISHERIES IN HARYANA

The economic impact of fisheries in Haryana encompasses multifaceted contributions to the state's GDP, employment generation, and the livelihoods of individuals engaged in this sector. A comprehensive understanding of these aspects necessitates an examination of fisheries production dynamics, distribution, and market trends within the state. Fisheries play a significant role in Haryana's economic development, not only through direct revenue generation but also by creating employment opportunities and enhancing livelihoods, particularly in rural areas where the sector often serves as a primary income source for many households. The contribution of fisheries to Haryana's GDP is a key indicator of its economic significance. Both capture fisheries and aquaculture contribute to the state's overall economic output. This contribution is measured in terms of the value of fish production, processing, distribution, and related activities. Analyzing the proportion of fisheries-related activities within the state GDP enables policymakers and economists to assess the sector's relative importance and its impact on Haryana's overall economy.

Employment generation is another critical aspect of the economic impact of fisheries in Haryana. The sector provides livelihoods for a diverse workforce, including fishers, fish farmers, processors, traders, transporters, and retailers. This employment spans various stages of the fisheries value chain, from fishing and aquaculture to processing, marketing, and sales. In rural areas, where the majority of fisheries activities are concentrated, employment in the sector is vital for poverty alleviation and rural development. By providing jobs and income opportunities, particularly to marginalized communities and small-scale fishers, the fisheries sector fosters inclusive growth and socioeconomic empowerment in Haryana.

Livelihoods in the fisheries sector extend beyond employment to encompass broader aspects of economic and social well-being. For many households in Haryana, especially those in riverine areas, fisheries are a primary source of income and sustenance. Fishermen and fish farmers often rely on traditional knowledge and skills passed down through generations. Additionally, fisheries support ancillary activities such as boat building, net making, ice production, and fish feed manufacturing, further enhancing livelihood opportunities for communities dependent on the sector. Livelihood diversification within the fisheries sector, including fish processing, value addition, and marketing, increases households' resilience to economic shocks and fluctuations in fish stocks.

The economic impact of fisheries in Haryana also includes trade and market dynamics. Fish and fishery products are traded domestically and internationally, with Haryana acting as both a producer and consumer market. The state's strategic location, proximity to major urban centers, and well-developed transportation infrastructure facilitate the distribution of fish products to markets within Haryana and neighboring states. The market for fish products in Haryana is diverse, encompassing fresh, frozen, dried, smoked, and value-added products catering to different consumer preferences and dietary needs. Market trends, price fluctuations, and consumer demand influence production decisions, investment patterns, and marketing strategies within the fisheries sector, shaping its economic impact on the state.

Government policies and interventions significantly shape the economic impact of fisheries in Haryana. Policy measures related to fisheries management, aquaculture development, trade regulations, infrastructure investment, and financial support influence the sector's growth, sustainability, and inclusivity. Investment in fisheries infrastructure, such as fishing harbors, landing centers, cold storage facilities, and market infrastructure, enhances the efficiency of fish production and distribution, benefiting both producers and consumers. Supportive policies for small-scale fishers, women in fisheries, and marginalized communities promote inclusive growth and poverty reduction within the sector.

Research and innovation also drive the economic impact of fisheries in Haryana by advancing technological improvements, productivity, and value addition. Scientific research in aquaculture technology, fish breeding, feed formulation, disease management, and post-harvest processing enhances the competitiveness and sustainability of the fisheries sector. Innovation in product development, packaging, branding, and marketing creates new market opportunities and increases the value of fish products, benefiting producers, processors, and consumers alike. Collaboration between government agencies, research institutions, industry stakeholders, and civil society organizations is essential for fostering a conducive environment for research and innovation in the fisheries sector.

Environmental sustainability is a critical consideration in assessing the economic impact of fisheries in Haryana. Sustainable fisheries management practices, including stock assessment, habitat conservation, bycatch reduction, and ecosystem-based management, are essential for maintaining the long-term viability of fish stocks and ecosystems. Addressing challenges such as overexploitation, habitat degradation, pollution, climate change, and invasive species requires a holistic approach that integrates ecological, social, and economic considerations, balancing the need for resource conservation with the imperative of livelihood security and economic development.

3.5 SUMMARY

The economic dynamics of fisheries in Haryana encompass a complex interplay of geographical, environmental, and socio-economic factors. Fisheries significantly contribute to the state's economy by providing employment opportunities, enhancing food security, and generating revenue through both domestic consumption and export. Despite being landlocked, Haryana has emerged as a key player in the fisheries sector, leveraging its aquaculture potential and innovative practices to drive economic growth and improve rural livelihoods. Examining the economics of fisheries in Haryana involves understanding various components, including production trends, market dynamics, government policies, and stakeholder challenges.

In recent years, fish production in Haryana has shown steady growth, spurred by investments in aquaculture and infrastructure. The state's diverse water bodies, such as ponds, reservoirs, canals, and rivers, offer ample opportunities for fish farming. Predominantly, carp species like Rohu, Catla, and Mrigal are cultivated, making up a significant portion of the state's fish production. The adoption of modern techniques such as pond lining, aeration, and water quality management has increased productivity and profitability. Furthermore, the introduction of high-yielding fish breeds and improved feed formulations has further enhanced production, positioning Haryana as a major contributor to the nation's fish output.

The economic importance of fisheries in Haryana extends beyond production to include employment generation and income augmentation, particularly in rural areas. Fish farming provides a livelihood for thousands of small and marginal farmers, offering an alternative income stream and mitigating agricultural risks. The decentralized nature of fish farming promotes widespread participation, empowering local communities and fostering socio-economic development. Additionally, the fisheries value chain, encompassing input supply, production, processing, marketing, and distribution, creates employment opportunities across various skill levels, from fish farmers and technicians to traders

CHAPTER 4

TYPES OF FISHES IN HARYANA

4.1 COMMON FISH SPECIES IN HARYANA

Haryana's ichthyofauna encompasses a diverse array of species adapted to its rivers, lakes, ponds, and reservoirs. This northern Indian state, characterized by its agricultural landscape and extensive water bodies, supports both native and introduced fish species, each contributing uniquely to the ecosystem and offering valuable resources for fisheries and aquaculture. A comprehensive understanding of Haryana's fish diversity involves examining their ecological roles, distribution patterns, and significance for biodiversity conservation and sustainable development. Native species in Haryana's aquatic habitats play a crucial role in maintaining ecological balance. These species have evolved to adapt to local conditions and occupy specific ecological niches. Native fishes, typically found in rivers and streams, contribute to nutrient cycling, food webs, and overall ecosystem health. For instance, the mahseer (Tor spp.), a large and iconic fish, is highly valued by anglers and serves as a flagship species for freshwater conservation in India.

Introduced species in Haryana, such as carp, catfish, and tilapia, have been introduced for aquaculture and recreational purposes. While these species can provide economic benefits and diversify local fisheries, they may also pose risks to native biodiversity. Introduced species can outcompete native ones for resources, disrupt food webs, and alter ecosystem dynamics. The Rohu (*Labeo rohita*) is a prominent freshwater fish in Haryana, belonging to the carp family (Cyprinidae). Widely distributed across India, Rohu is a popular food fish known for its taste and nutritional content. It is an omnivorous species, feeding on algae, aquatic plants, insects, and small crustaceans, and is extensively farmed in Haryana's aquaculture systems.

Similarly, the Catla (*Catla catla*), another major carp species, is significant in Haryana's aquaculture. This herbivorous fish primarily consumes phytoplankton and aquatic plants. Catla's fast growth rate and high market demand make it a staple in integrated polyculture systems, where it is farmed alongside other carp species like Rohu and Mrigal (*Cirrhinus cirrhosus*). The Grass Carp (*Ctenopharyngodon idella*), introduced for weed control and aquaculture, is notable for its herbivorous diet and ability to manage aquatic vegetation. However, its introduction can lead to ecological imbalances by disrupting native plant communities and altering habitat structures.

Silver Carp (*Hypophthalmichthys molitrix*), another introduced species, is valued for its fast growth and high protein content. This filter-feeding fish primarily consumes phytoplankton and suspended organic matter, often raised in aquaculture ponds. Nonetheless, its potential impact on native ecosystems through competition for food and habitat remains a concern. Haryana also hosts catfish species such as the Walking Catfish (*Clarias batrachus*) and African Catfish (*Clarias gariepinus*). Known for their whisker-like barbels and nocturnal feeding behavior, these bottom-dwelling fishes are popular in aquaculture but can become invasive if they establish wild populations. Tilapia (Oreochromis spp.), another non-native species, is prevalent in Haryana's aquaculture due to its fast growth, high reproductive capacity, and environmental tolerance. While beneficial for aquaculture, Tilapia can become invasive, threatening local ecosystems.

4.2 PRESENCE IN DIFFERENT DISTRICTS

The geographical distribution of fish species across various administrative regions or districts within a country is critical to understanding their ecology and informing conservation efforts. Fish species exhibit diverse distribution patterns influenced by habitat availability, environmental conditions, historical events, and human activities. Grasping the presence of fish in different districts is vital for effective management and conservation, offering insights into species diversity, population dynamics, and ecosystem health. Habitat preferences and ecological requirements predominantly shape the distribution of fish species. Freshwater species, for instance, are typically found in rivers, streams, and lakes, where factors such as water flow, temperature, depth, and substrate composition play crucial roles. Certain species may be adapted to specific habitat features, including rocky substrates, sandy bottoms, submerged vegetation, or fast-flowing currents.

Local adaptations to environmental conditions further influence distribution patterns. Fish exhibit remarkable phenotypic and behavioral adaptations, enabling them to exploit ecological niches and thrive in diverse environments. These adaptations include physiological traits for osmoregulation, thermoregulation, and oxygen uptake, as well as morphological features for feeding, locomotion, and predator avoidance. For example, fish in high-altitude streams may be adapted to low oxygen levels and rapid currents, while tropical species may possess specialized morphologies for navigating dense vegetation. Historical events and evolutionary processes also shape fish distribution. Geological events such as continental drift, tectonic movements, and glacial cycles have altered aquatic habitats and isolated populations over geological timescales. These processes contribute to the divergence of fish populations and the formation of distinct species assemblages. Evolutionary mechanisms, including speciation, hybridization, and adaptive radiation, further enhance species diversity and distribution patterns.

Human activities significantly impact fish distribution. Habitat modification, pollution, overfishing, invasive species introduction, and climate change are major drivers of change in aquatic ecosystems. Habitat loss and degradation from deforestation, urbanization, agriculture, and dam construction fragment aquatic habitats, isolate populations, and disrupt migration routes, leading to population declines and biodiversity loss. Pollution impairs water quality, affecting fish health and reproduction. Overfishing depletes stocks and alters marine food webs, while bycatch poses additional threats to vulnerable species. Invasive species outcompete natives and disrupt community dynamics. Climate change alters habitat suitability and species ranges through shifts in temperature, precipitation, sea level, and ocean currents.

Adaptation to local environmental conditions is essential for fish species to survive and thrive across districts. Genetic changes and phenotypic plasticity enable fish to adjust their morphology, physiology, and behavior in response to environmental cues. For instance, fish may exhibit plastic responses in body size, coloration, feeding preferences, and reproductive timing across different habitats or seasons.

4.3 FISH PRODUCTION STATISTICS

The trajectory of fish production in Haryana has markedly ascended, catalyzing a transformation from a predominantly agricultural landscape to one that increasingly acknowledges the vast potential of aquaculture. Situated in northern India, Haryana has adeptly utilized its varied water resources-encompassing rivers, reservoirs, canals, and ponds-to enhance fish production. This progression has been propelled by a synergistic blend of strategic initiatives, technological advancements, and robust governmental support, positioning the state as an emerging contributor to India's fisheries sector.

Historically, Haryana's geographical and climatic conditions posed challenges to the development of fish farming. However, growing awareness of the nutritional and economic benefits of fish, coupled with the imperative for agricultural diversification, has elevated fish farming to a priority status. The Haryana government, recognizing the sector's potential, has introduced a range of schemes and incentives to foster fish farming among local agrarians.

Fish production in Haryana predominantly focuses on freshwater species, due to the state's lack of marine resources. Key species farmed include Indian major carps such as Catla (*Catla catla*), Rohu (*Labeo rohita*), and Mrigal (*Cirrhinus mrigala*), valued for their rapid growth rates, market demand, and adaptability to local environmental conditions. Additionally, there is a growing interest in exotic species like Common Carp (*Cyprinus carpio*), Grass Carp (*Ctenopharyngodon idella*), and Silver Carp (*Hypophthalmichthys molitrix*), known for their accelerated growth and enhanced productivity within controlled aquaculture systems.

Statistical data indicates a significant upward trend in Haryana's fish production. Over the past decade, the area dedicated to fish farming has expanded considerably, driven by the conversion of low-lying, waterlogged, and saline-affected agricultural lands into productive fish ponds. This innovative land use has dual benefits: improving farmers' livelihoods and boosting the state's fish production. Currently, Haryana has approximately 20,000 hectares devoted to fish farming, yielding around 1.5 lakh tonnes of fish annually. This marks a substantial increase from previous years, underscoring the efficacy of aquaculture development programs.

The adoption of scientific fish farming practices has been a pivotal factor in this growth. The Haryana government, in partnership with agricultural universities and research institutions, has promoted modern aquaculture techniques. These include the cultivation of high-yielding fish varieties, enhanced feed and nutrition management, disease control measures, and the implementation of polyculture systems, where multiple fish species are co-farmed to maximize resource utilization and productivity.

Further bolstering this growth is the state's investment in robust infrastructure to support fish farming operations. Establishing hatcheries and seed production centers throughout Haryana ensures a consistent supply of quality fish seeds, essential for sustaining high production levels. These centers annually produce millions of fish fry and fingerlings, distributed to farmers at subsidized rates. This strategy not only lowers the initial investment costs for farmers but also guarantees healthy and genetically superior fish stocks, reinforcing the sustainability and growth of Haryana's fish production industry.

4.4 FISHERIES' CONTRIBUTION TO HARYANA'S GDP

Fisheries constitute a notable economic pillar in Haryana, a state traditionally recognized for its agricultural and industrial sectors. Despite its inland location devoid of coastal advantages, Haryana has strategically invested in fisheries, tapping into its water resources for economic sustenance. The economic ramifications of fisheries extend across several dimensions, encompassing aquaculture, fish production, employment, and revenue generation. Aquaculture stands as a pivotal component within Haryana's fisheries framework, representing a concerted effort to cultivate fish and aquatic organisms within controlled environments such as ponds and reservoirs. The state's proactive measures include substantial investments, subsidies, and infrastructural development aimed at fostering a conducive environment for fish farming. These initiatives have yielded tangible results, evidenced by escalated fish production and enhanced incomes among fish farmers, thereby contributing positively to the state's GDP.

The fish output in Haryana serves both domestic consumption and export markets, amplifying its economic significance. The state boasts a diverse array of freshwater species, including carp, catfish, tilapia, and murrel, catering to local demand and beyond. This localized production not only meets regional protein requirements but also diminishes reliance on imported seafood. Furthermore, Haryana's active participation in export markets underscores its capacity to supply high-quality fish products internationally, thereby augmenting foreign exchange reserves. A critical facet of the fisheries sector lies in its role as a significant employer, offering livelihood opportunities to a substantial segment of the populace. From fish farmers to processors, traders, and ancillary service providers, the industry engenders employment across various strata, particularly in rural areas. This employment expansion not only addresses livelihood challenges but also fosters ancillary growth in related industries, contributing to broader socio-economic development.

Revenue generation constitutes another vital dimension of Haryana's fisheries sector, manifested through taxation, fees, and export earnings. Governmental levies on fish production and trade, coupled with revenues from export ventures, bolster state coffers. Additionally, the state's engagement in domestic and international fish markets facilitates value addition and price premiumization, augmenting overall revenue streams from the fisheries sector. Comparative analysis with coastal counterparts sheds light on Haryana's unique position and potential within the fisheries landscape. While lacking direct access to marine resources, the state's emphasis on inland fisheries underscores efficient land and water resource utilization. Despite coastal states' inherent advantage in total fish production, Haryana's per capita consumption and GDP contribution per unit of resource utilization remain competitive, if not superior. In terms of aquaculture practices, Haryana demonstrates a commitment to modernization and efficiency, mitigating challenges such as coastal erosion and pollution faced by coastal states. The state's inland aquaculture systems offer stability and resilience, bolstered by advantageous factors like biosecurity and water quality management. Such factors, coupled with Haryana's socio-economic inclusivity, underscore the sector's broader impact on poverty alleviation and rural prosperity.

4.5 SUMMARY

In Haryana, the aquaculture industry demonstrates the state's commitment to fisheries management and economic development. Despite its landlocked status, Haryana sustains a thriving aquaculture sector, utilizing reservoirs, ponds, canals, and tanks. Government initiatives aimed at promoting fish farming have significantly boosted local economy and food security. Indian major carp species like rohu (*Labeo rohita*), catla (*Catla catla*), and mrigal (*Cirrhinus mrigala*) dominate Haryana's aquaculture scene due to their rapid growth, high demand, and adaptability. Common carp (*Cyprinus carpio*) is another economically important species, favored for its hardiness and fast growth. Exotic carp species such as grass carp (*Ctenopharyngodon idella*), silver carp (*Hypophthalmichthys molitrix*), and bighead carp (Hypophthalmichthys nobilis) are also cultivated, offering efficient feed conversion and aquatic weed control.

Freshwater prawns like Macrobrachium rosenbergii and M. malcolmsonii have gained popularity for their export potential and adaptability to local conditions. Ornamental fish farming is emerging as a lucrative venture, driven by both domestic and international demand for colorful species like goldfish, guppies, and tetras. Fish farming in Haryana contributes to employment generation, poverty alleviation, and rural development, benefiting small-scale farmers and marginalized communities. Supporting industries such as feed manufacturing and aquaculture equipment supply further stimulate economic growth. However, challenges such as limited access to quality seed and feed, inadequate infrastructure, water scarcity, and market fluctuations hinder sustainable growth. To address these challenges, investment in research and development, capacity building, and market development is essential.

CHAPTER 5

FISHERIES INSTITUTIONS IN HARYANA

5.1 GOVERNMENT INSTITUTIONS

In the verdant expanse of Haryana, a state nestled in the northern reaches of India, fisheries institutions stand as guardians of aquatic ecosystems, stewards of sustainable practices, and architects of economic vitality. Collaborative in nature, these institutions, both governmental and non-governmental, converge their efforts to orchestrate a symphony of management, regulation, and advocacy within the fisheries domain. Spearheading this concerted endeavor is the Haryana Fisheries Department, a bastion of governmental oversight and strategic planning. Operating under the auspices of the Department of Fisheries, Government of Haryana, and helmed by the Director of Fisheries, this pivotal entity is charged with a multifaceted mandate encompassing policy formulation, regulatory enforcement, and provision of advisory services.

Central to the mission of the Haryana Fisheries Department is the promotion of aquaculture as a linchpin of rural livelihoods and economic prosperity. With a suite of schemes and programs under its aegis, including the Rashtriya Krishi Vikas Yojana (RKVY) and the National Fisheries Development Board (NFDB), the department extends a helping hand to fish farmers, offering financial incentives, technical expertise, and capacity-building initiatives. Concurrently, efforts are directed towards the conservation and sustainable management of natural fisheries resources, facilitated through collaborative engagements with governmental agencies, research bodies, and local communities.

Complementary to the operational framework of the Haryana Fisheries Department are legislative instruments delineating the rights, responsibilities, and regulatory parameters governing fisheries activities within the state. Chief among these is the Haryana Fisheries Act, buttressed by the Haryana Aquaculture Rules, which collectively furnish the legal scaffolding for the enforcement of standards, licensing protocols, and environmental safeguards across the aquaculture spectrum. Embedded within these statutory frameworks are provisions aimed at curtailing illegal fishing practices, preserving fish habitat integrity, and adjudicating disputes pertinent to fisheries governance.

Aligned with these statutory imperatives is the Haryana Fisheries Policy, a strategic roadmap delineating the state's vision for the sustainable evolution of fisheries and aquaculture enterprises. Rooted in principles of scientific rigor, community empowerment, and private sector engagement, this policy blueprint accentuates the imperative of gender inclusivity, environmental stewardship, and technological innovation in propelling the fisheries sector towards a trajectory of inclusive growth and resilience.

5.2 RESEARCH AND EDUCATIONAL INSTITUTIONS

In Haryana, the nexus of research and educational institutions is instrumental in sculpting the fisheries landscape within the state, serving as fulcrums for the cultivation of knowledge, skills, and technologies imperative for the sustainable management of fisheries and aquaculture practices. Foremost among these institutions stands the Department of Fisheries, operating under the auspices of the state government. Entrusted with the formulation of policies, execution of programs, and provision of technical expertise, the Department spearheads endeavors aimed at fostering the advancement of the fisheries sector in Haryana. Through a myriad of initiatives, the department endeavors to augment fish production, refine fish quality, and ameliorate the socio-economic fabric of fishers and fish farmers statewide.

Beyond governmental entities, academic bastions such as universities and colleges significantly bolster the fisheries domain in Haryana. These institutions proffer specialized curricula, training regimens, and research platforms in fisheries science, aquaculture, and allied fields. Eminent among these is Maharshi Dayanand University in Rohtak and Chaudhary Charan Singh Haryana Agricultural University in Hisar, renowned for their comprehensive agricultural and fisheries science programs. Here, students are imbued with comprehensive education and practical skills, priming them for impactful roles in fisheries and aquaculture domains. Moreover, these academic institutions engage in pioneering research endeavors targeting pivotal challenges confronting the fisheries sector in Haryana, including water quality management, disease mitigation, and sustainable aquaculture methodologies.

Moreover, research institutes and centers dedicated to fisheries science and aquaculture emerge as pivotal catalysts in propelling knowledge and innovation in Haryana. These entities orchestrate applied research, technology incubation, and extension endeavors to bolster the growth of the fisheries sector while enhancing the livelihoods of fishers and fish farmers. Notably, the regional center of the Central Institute of Fisheries Education (CIFE) in Rohtak is pivotal, concentrating on freshwater aquaculture research and training. Collaborating closely with local stakeholders, this center endeavors to devise and disseminate technologies aimed at augmenting fish production and productivity in Haryana's aquatic habitats.

5.3 COOPERATIVE SOCIETIES AND NGOS

In the multifaceted fisheries landscape of Haryana, cooperative societies and Non-Governmental Organizations (NGOs) stand as pivotal forces, wielding significant influence over the development, management, and sustainability of the sector. These entities assume diverse roles, ranging from providing crucial support to fishers and fish farmers to spearheading conservation initiatives and advocating for policy reforms. In Haryana, cooperative societies and NGOs have emerged as indispensable stakeholders, driving socio-economic progress and environmental stewardship within the fisheries domain. At the grassroots level, cooperative societies in Haryana play a central role in mobilizing and empowering fishers and fish farmers. Often formed by local communities or collectives of fishers, these societies serve as platforms for collective decision-making and action. Through collaborative efforts and resource pooling, cooperative societies facilitate access to vital inputs such as seeds, feeds, and equipment at competitive rates, thereby amplifying productivity and profitability among members. Additionally, these societies streamline the marketing and distribution of fish products, enabling fishers to secure better prices and improve their economic prospects.

Moreover, cooperative societies foster social cohesion and mutual aid within fisher communities in Haryana. Implementing initiatives like savings and credit schemes, these societies offer financial assistance to members during times of need, mitigating socio-economic vulnerabilities. Furthermore, through workshops, training programs, and awareness campaigns, cooperative societies equip fishers and fish farmers with essential knowledge and skills in fisheries management, aquaculture techniques, and market dynamics, thereby enhancing their capacities and resilience. In tandem with cooperative efforts, NGOs in Haryana advocate for the rights and welfare of fishers while promoting sustainable fisheries practices. Driven by principles of social equity and environmental conservation, these organizations undertake diverse activities aimed at empowering local communities and safeguarding natural resources.

Central to their mission is the dissemination of knowledge on sustainable fisheries management and the conservation of aquatic ecosystems. Through community engagement initiatives, educational endeavors, and stakeholder consultations, NGOs cultivate a culture of environmental stewardship and responsible fishing practices among fishers and allied stakeholders.

Furthermore, NGOs collaborate extensively with governmental bodies, academic institutions, and international partners to implement projects fostering sustainable fisheries and inclusive development. These collaborations span research endeavors, monitoring activities, and conservation interventions aimed at assessing fish stocks, biodiversity, and habitat integrity. Leveraging their expertise and networks, NGOs provide valuable insights and technical support to bolster fisheries management initiatives in Haryana.

5.4 SUMMARY

In Haryana, the fisheries sector thrives within a comprehensive institutional framework, uniting governmental and non-governmental entities in a collective effort to foster sustainable aquaculture practices, bolster fish production, and uplift the livelihoods of fishers and fish farmers. Spearheaded by the Department of Fisheries, governance encompasses policy formulation, program implementation, and regulatory oversight, supported by divisional structures addressing aquaculture development, fish seed production, health management, and extension services. Collaboration with governmental bodies, research institutions, and industry stakeholders underscores efforts to address sectoral challenges and promote coordinated action.

At the heart of this framework lies a concerted push for aquaculture development, with the Department of Fisheries incentivizing modern practices among fish farmers through subsidies, technical support, and educational outreach. Simultaneously, a robust fish seed production infrastructure, comprising both government-operated facilities and private hatcheries, ensures the availability of certified stock while adhering to stringent quality control measures. Disease surveillance and management, buttressed by research collaborations and veterinary expertise, further fortify the sector against health threats. Extension services emerge as a linchpin for knowledge dissemination and capacity building, with tailored training initiatives empowering stakeholders across the fisheries value chain. Research and development efforts, facilitated by strategic partnerships with national and regional institutes, drive innovation in breeding, nutrition, water management, and market strategies, enhancing sectoral competitiveness and sustainability. Concurrently, marketing and value chain interventions, coupled with investments in infrastructure and human capital, bolster market access and economic resilience among fisher communities.

CHAPTER 6

CHALLENGES OF FISHERIES FARMING IN HARYANA

6.1 ENVIRONMENTAL CHALLENGES

Fisheries farming in Haryana encounters diverse challenges, spanning environmental, socioeconomic, and logistical realms, which profoundly influence the sustainability and success of operations in the region. Environmental issues, particularly water quality concerns, significantly impede fish cultivation, affecting both aquatic ecosystem health and fish productivity. Inadequate water quality, marked by elevated pollutants, turbidity, and low oxygen levels, detrimentally affects fish health and growth. Pollution from sources like agricultural runoff and industrial discharge introduces harmful substances, posing risks to fish populations and human health, thus compromising the sustainability of aquaculture. Additionally, climate change exacerbates these challenges, disrupting habitats and exacerbating water quality issues, further compromising fish farming systems' productivity and resilience.

Water scarcity emerges as another critical challenge, especially amidst rising freshwater demands and competition from other sectors. Limited freshwater availability restricts fish farming expansion and constrains suitable pond construction sites, exacerbating water quality concerns in aquaculture ponds. Addressing water scarcity necessitates innovative water management strategies to optimize water use and minimize wastage. Inadequate infrastructure and technical expertise also impede progress, hindering access to essential resources and training. Market access and value chain constraints pose additional obstacles, limiting fish farmers' ability to access lucrative markets and obtain fair prices, thus impacting profitability and economic opportunities. Regulatory and policy challenges, including outdated regulations and bureaucratic barriers, further complicate the regulatory landscape for fish farming operations, leading to compliance challenges and legal risks. Social and cultural factors also play a significant role, influencing community attitudes towards sustainable aquaculture practices.

6.2 ECONOMIC AND MARKET CHALLENGES

Haryana, a prominent player in India's agricultural sector, faces limitations in its fisheries industry despite its potential. Here, we examine the key economic and market challenges that impede the growth and development of this sector. Limited market access is a primary constraint for Haryanvi fish farmers. The state lacks a well-developed network of dedicated fish markets and cold chain infrastructure. This results in substantial post-harvest losses and diminished profitability. Traditional marketing channels often involve multiple intermediaries, reducing the producer's share of the final price. Furthermore, the absence of proper market information systems leaves producers uninformed about prevailing market prices, making them susceptible to exploitation by middlemen.

Inadequate transportation and storage infrastructure for fish further exacerbates the situation. Limited access to refrigerated trucks and cold storage facilities leads to significant spoilage, particularly during peak harvest seasons. This not only reduces the overall production reaching consumers but also discourages investment in intensive aquaculture practices. Additionally, poor road connectivity in some rural areas hinders fish farmers from transporting their produce to major markets, further restricting their reach.

The Haryanvi fisheries sector experiences intense competition from both domestic and international sources. States like Andhra Pradesh and

Gujarat, with established market linkages and well-developed aquaculture infrastructure, often flood the market with cheaper fish, putting pressure on prices received by Haryanvi producers. Increased competition also comes from cheaper imports of fish and fishery products from Southeast Asia. Price volatility presents another major challenge. Fish prices are highly susceptible to fluctuations in supply and demand, making it difficult for fish farmers to plan their production cycles and secure stable incomes. Factors like seasonal variations in fish availability, sudden changes in consumer preferences, and disruptions in transportation networks can all contribute to price volatility. This uncertainty discourages investment in modern aquaculture techniques and deters potential new entrants to the sector.

The cost of essential inputs like fish seed, feed, and fertilizers has been steadily rising in recent years. These increases are often driven by factors beyond the control of fish farmers, such as fluctuations in global commodity prices and disruptions in supply chains. High input costs can significantly erode profit margins, especially for small-scale farmers with limited resources. Biosecurity concerns pose a significant threat to the sustainability of Haryana's fisheries sector. Disease outbreaks can devastate fish stocks and cause severe economic losses for farmers. Limited access to diagnostic facilities and veterinary services makes it difficult for farmers to detect and respond to disease outbreaks promptly. Additionally, the use of poor-quality or contaminated fish seed can introduce pathogens into aquaculture systems, further exacerbating biosecurity risks.

The existing policy and regulatory frameworks for fisheries in Haryana may not be adequately tailored to address the specific needs of the sector. Complex licensing procedures and bureaucratic hurdles can discourage investment and hinder the adoption of new technologies. Furthermore, a lack of clarity on water use rights for aquaculture purposes can create uncertainty for fish farmers, particularly those operating in water-scarce regions. Inadequate extension services and limited access to training programs further restrict the development of the sector. Fish farmers often lack the knowledge and skills necessary to implement best aquaculture practices, manage resources efficiently, and adopt new technologies. An ineffective extension system hinders the dissemination of vital information on disease prevention, water quality management, and sustainable aquaculture practices.

Access to finance is a major hurdle for Haryanvi fish farmers, particularly small-scale operators. Commercial banks and financial institutions often perceive aquaculture as a high-risk sector and are hesitant to provide loans. The lack of access to affordable credit restricts investment in pond infrastructure, stocking quality fish seed, and adopting new technologies. Limited access to risk management options further weakens the financial resilience of fish farmers. The availability of insurance products specifically designed for the fisheries sector is scarce in Haryana. Without proper risk management tools, fish farmers are left vulnerable to sudden price drops, disease outbreaks, and other unforeseen events that can cause significant financial losses.

6.3 SOCIAL AND INSTITUTIONAL CHALLENGES

Beyond economic and market constraints, social and institutional challenges form a complex web that impedes the development and inclusivity of Haryana's fisheries sector. Addressing these issues is critical for establishing a sustainable and equitable fisheries system that empowers all stakeholders. The entrenched caste system in Haryana restricts entry for certain communities into fish farming. Upper castes often control access to land and water resources, limiting opportunities for marginalized communities to participate in aquaculture. This social exclusion not only hinders overall sectoral growth but also exacerbates existing social inequalities.

Women play a vital role in post-harvest activities like fish processing, sorting, and marketing. However, they are frequently excluded from decision-making processes, lack access to training and resources, and receive lower wages compared to men. Empowering women in fisheries through capacity building programs, financial inclusion initiatives, and promoting their participation in fish farmer cooperatives can enhance efficiency, improve livelihoods, and foster social equity within the sector. The fisheries sector in Haryana, particularly shrimp aquaculture, often relies heavily on migrant labor. These workers, especially those from marginalized communities, are vulnerable to exploitation due to the absence of formal contracts, poor working conditions, and inadequate social security measures. Ensuring fair labor practices, promoting adherence to minimum wages, and providing basic social security benefits are essential for protecting the rights of migrant workers and fostering a sustainable and ethical fisheries sector.

Many fish farmers in Haryana, particularly small-scale operators, lack the necessary knowledge and skills to adopt best aquaculture practices. Traditional practices often persist, leading to inefficiencies in resource management, lower productivity, and increased environmental impacts. Investing in training programs on pond management, disease prevention, and sustainable aquaculture techniques is crucial for empowering fish farmers to improve their productivity and profitability. The current institutional framework for fisheries development in Haryana faces challenges. Inadequate staffing, limited resources, and bureaucratic hurdles can hinder the effectiveness of government fisheries departments. Strengthening extension services is critical for disseminating knowledge on best practices, new technologies, and government schemes to fish farmers. A well-equipped and well-staffed extension network can play a pivotal role in bridging the information gap between researchers and farmers, facilitating innovation adoption, and promoting sustainable aquaculture practices. Small-scale fish farmers often face difficulties accessing credit from formal financial institutions due to the perceived high risks associated with aquaculture and a lack of collateral. This restricts their ability to invest in pond infrastructure, quality fish seed, and new technologies necessary to improve productivity and compete effectively in the market. Developing targeted credit schemes with relaxed collateral requirements and exploring alternative financing models like microfinance institutions can improve financial inclusion and empower small-scale fish farmers to invest in growth and modernization.

6.4 TECHNOLOGICAL CHALLENGES

Haryana's fisheries sector holds significant potential, yet faces limitations in adopting advanced aquaculture technologies. This hinders modernization, efficiency, and overall productivity. Addressing these technological challenges is critical for promoting sustainable practices, enhancing competitiveness, and ensuring the sector's long-term viability. The widespread adoption of advanced aquaculture technologies remains low in Haryana, particularly among small-scale farmers. Traditional practices persist, characterized by low stocking densities, limited aeration, and reliance on natural productivity. This reluctance to embrace new technologies stems from several factors:

Knowledge Deficits: Many fish farmers lack awareness of the benefits and functionalities of advanced systems like biofloc, recirculating aquaculture systems (RAS), and biofloc-RAS. Inadequate extension services and limited access to training programs further hinder knowledge dissemination and technology adoption.

Perceived Cost Constraints: The initial investment costs associated with advanced technologies can be perceived as high by resource-limited small-scale farmers. The absence of readily available financing options and targeted subsidies for technology adoption further restricts accessibility.

Skill Gaps and Technical Expertise: Implementing and operating advanced technologies often require specialized skills. The current skillset of many fish farmers may not be readily adaptable, necessitating targeted training programs to bridge the gap.

High-quality and readily available fish seed are crucial for successful aquaculture operations. However, the fisheries sector in Haryana faces challenges related to broodstock management and seed quality:

Limited Access to High-Quality Broodstock: The availability of genetically improved and disease-resistant broodstock for commercially important fish species is limited. Reliance on traditional seed production methods often leads to inbreeding and reduced fish quality.

Biosecurity Concerns and Disease Outbreaks: Inadequate biosecurity measures in broodstock management and seed production facilities can increase the risk of disease outbreaks. The lack of readily available diagnostic tools and limited access to veterinary services further complicate disease prevention and control efforts.

Dependence on Private Hatcheries: Many fish farmers rely on private hatcheries for fish seed, which can be of variable quality and susceptible to diseases. Strengthening government hatcheries and promoting certified seed producers can ensure the availability of high-quality seed.

Feed costs often constitute a significant portion of operational expenses in aquaculture. However, limitations exist in terms of feed management practices in Haryana:

Overreliance on Traditional Feeds: Many fish farmers continue to rely on traditional feeds like groundnut cake and rice bran, which may not be nutritionally balanced to meet the specific requirements of different fish species and life stages. This can lead to inefficiencies in feed utilization and increased production costs.

Limited Availability of Specialized Feeds: The availability of commercially produced, species-specific, and nutritionally balanced feeds may be limited in certain regions of Haryana. This restricts the ability of fish farmers to optimize growth rates and improve feed conversion ratios.

Knowledge Gaps in Feed Management: Fish farmers may lack adequate knowledge about proper feed management practices, such as calculating feed rations based on fish size and water quality parameters. This can lead to overfeeding or underfeeding, impacting fish health and productivity.

Maintaining optimal water quality is essential for healthy fish growth and preventing disease outbreaks. However, water quality management presents challenges in Haryana:

Limited Adoption of Water Treatment Technologies: The adoption of water treatment technologies like aeration systems, biofloc technology, and biofiltration systems is often low, particularly among small-scale farmers. This can lead to fluctuations in water quality parameters like dissolved oxygen and ammonia levels, impacting fish health.

Water Scarcity and Competition for Resources: Water scarcity is a growing concern in Haryana, leading to competition for water resources between agriculture and aquaculture. Inadequate infrastructure for water storage and limited adoption of water reuse techniques can further exacerbate water scarcity challenges.

Environmental Concerns and Pollution Risks: The use of excessive antibiotics and chemicals in aquaculture can contribute to antimicrobial resistance and environmental pollution. Promoting environment-friendly aquaculture practices, such as biofloc technology and integrated aquaculture with agriculture, is crucial for mitigating these risks.

Inadequate investment in research and development (R&D) can hinder the development and adoption of new and improved aquaculture technologies in Haryana. Focusing R&D efforts on areas such as: Species improvement and selective breeding programs: Developing genetically improved fish breeds with faster growth rates, disease resistance, and improved tolerance to environmental conditions. Development of cost-effective and efficient aquaculture technologies: Researching and developing cost-effective technologies suitable for small-scale farmers.

6.5 SUMMARY

Despite efforts to promote fisheries farming in Haryana, its development faces significant challenges. Environmental constraints like limited water resources due to Haryana's climate and competition from other sectors restrict suitable land and water availability. Additionally, compromised water quality from pollution and over-extraction further hinders aquaculture. Technologically, reliance on traditional practices limits productivity and efficiency. While modern technologies like recirculating aquaculture systems can optimize production, high costs and inadequate support services like training and technical assistance hinder their adoption. Regulatory and policy shortcomings also impede growth. The lack of clear frameworks creates uncertainty for investment, while weak governance structures enable illegal and This includes promoting sustainable land and water management practices, investing in research, capacity building, and technology transfer initiatives. Strengthening regulations, empowering fish farmers, and fostering gender equality, social inclusion, and community participation are all crucial for creating a sustainable and equitable fisheries sector in Haryana.

CHAPTER 7

RELEVANCE OF WETLANDS AND SCOPE OF FISHERIES IN HARYANA

7.1 IMPORTANCE OF WETLANDS

Wetlands, with their vibrant tapestry of life, represent some of Earth's most crucial ecosystems. In Haryana, a landlocked state in northern India, their significance transcends their inherent value, extending to socio-economic benefits and potential for fisheries development. Understanding these ecosystems' functions and their role in supporting biodiversity is paramount for effective conservation and sustainable management. Defined by water's dominance in shaping their environment, wetlands encompass a diverse array of habitats like marshes, swamps, and floodplains. In Haryana, these ecosystems play a critical role in water resource provision, biodiversity support, and mitigating floods and droughts. Despite occupying a small portion of the state's land area, wetlands exert a disproportionate influence on ecological balance and livelihoods.

Wetlands' ecological functions are multifaceted, contributing to local, regional, and global ecosystem health. One key function is water storage and regulation. Acting as natural sponges, they absorb excess water during heavy rainfall, releasing it slowly during dry periods, mitigating floods and droughts. In Haryana, wetlands regulate river and stream flows, reducing downstream flood risks and replenishing groundwater supplies. Water purification and nutrient cycling are other vital functions. The intricate network of plants, microbes, and other organisms within wetlands filters pollutants from water, improving quality and reducing the impact of agricultural runoff and industrial discharge. Additionally, wetlands trap and store nutrients like nitrogen and phosphorus, preventing them from entering water bodies and causing eutrophication, a process detrimental to aquatic life. In Haryana, wetlands act as natural filters, enhancing the quality of water in rivers, lakes, and reservoirs, ultimately supporting aquatic biodiversity.

Wetlands' diverse combination of water, nutrients, and habitat types creates biodiversity hotspots. They provide habitat for a vast array of plants and animals, including migratory birds, amphibians, fish, and invertebrates. In Haryana, wetlands serve as crucial stopovers for migratory birds along the Central Asian Flyway, offering vital feeding and resting areas during their long journeys. Additionally, these ecosystems support resident bird species like waterfowl, herons, and raptors, making them popular destinations for birdwatchers and ecotourists.

Conservation and sustainable management of wetlands can significantly benefit fisheries development in Haryana. They provide crucial breeding, feeding, and nursery grounds for many fish species, influencing their reproductive success and population dynamics. The rich biodiversity of wetlands supports a diverse fish community, including native and migratory species, offering ample opportunities for fisheries development and aquaculture. By preserving and restoring wetlands, Haryana can enhance its fisheries resources and promote sustainable livelihoods for fishing communities.

The value of wetlands extends beyond their ecological and socio-economic importance to encompass cultural and recreational significance. These ecosystems have long been integral to human societies, providing food, water, and materials. In Haryana, wetlands are interwoven with cultural practices, traditions, and festivals, reflecting their deep-rooted connection to local communities. They also offer recreational opportunities for nature enthusiasts, photographers, and outdoor enthusiasts, providing space for birdwatching, boating, and nature trails.

Despite their immense value, wetlands face numerous threats from human activities – urbanization, agriculture, pollution, and climate change. Conversion of wetlands for agriculture, infrastructure development, and urban expansion leads to their loss and degradation, resulting in habitat fragmentation, biodiversity loss, and disruption of ecological processes. Industrial discharge, agricultural runoff, and solid waste pollute wetland waters, threatening the health of aquatic ecosystems and the species that depend on them.

7.2 WETLANDS IN HARYANA

Wetlands in Haryana weave a vital thread into the state's ecological tapestry, playing a critical role in biodiversity conservation, water management, and socio-economic development. These diverse ecosystems encompass a spectrum of habitats, from marshlands and swamps to permanent lakes and flowing rivers, each harboring unique flora and fauna adapted to the wetland environment. Distributed across the state, these crucial sanctuaries provide essential services like flood control, groundwater recharge, and water purification, while acting as havens for resident and migratory bird populations.

Sultanpur National Park, a renowned bird sanctuary and Ramsar site located in Gurugram district, exemplifies Haryana's wetland significance. This mosaic of wetlands, grasslands, and woodlands offers refuge to over 250 bird species, including migratory visitors from Central Asia, Europe, and Siberia. The park's wetlands are vital for these avian populations, providing vital feeding, roosting, and breeding grounds amidst the surrounding urban landscape. Furthermore, Sultanpur serves as a popular destination for birdwatchers and nature enthusiasts, contributing to ecotourism and environmental education in the region.

Another significant wetland is the Bhindawas Wildlife Sanctuary in Jhajjar district. This man-made reservoir, created by damming the Jui Canal, has evolved into a sprawling wetland complex encompassing marshes, ponds, and grasslands. The sanctuary teems with resident and migratory birdlife, including waterfowl, waders, and raptors. Species like the sarus crane, bar-headed goose, and northern pintail are frequent winter visitors. Beyond its avian diversity, Bhindawas harbors a variety of fish, amphibians, and reptiles, contributing to the sanctuary's overall ecological value. The wetland plays a crucial role in groundwater recharge and flood regulation, acting as a natural reservoir during monsoons and mitigating the impacts of droughts and water scarcity.

Haryana's wetland network extends beyond designated protected areas, encompassing a variety of natural and man-made habitats across the state. Wetlands can be found woven into diverse landscapes, including river floodplains, agricultural fields, and even urban areas, each offering unique ecological functions and services. For instance, the Yamuna river basin supports extensive wetlands along its course through Haryana, providing habitat for waterfowl, fish, and aquatic plants. These wetlands are critical for maintaining water quality, supporting biodiversity, and sustaining livelihoods for local communities dependent on fishing and agriculture.

Man-made wetlands, such as irrigation tanks, reservoirs, and waterlogged areas, also play a significant role in Haryana's water story. These wetlands, created for various purposes including agriculture, water supply, and wastewater treatment, may lack the ecological diversity of natural wetlands, yet they still contribute significantly to water management and biodiversity conservation. For example, waterlogged areas in Haryana's rice-growing regions provide crucial habitat for wetland birds and aquatic species during the monsoon season when fields are flooded for paddy cultivation. The distribution of wetlands across Haryana reflects the state's diverse physiographic and hydrological features. Found in both the plains and the foothills of the Shivalik range, these ecosystems exhibit varying degrees of permanence and ecological characteristics. In western districts like Sirsa and Fatehabad, wetlands are often linked to canal networks and agricultural fields, providing habitat for waterfowl and supporting irrigated agriculture. Eastern districts like Kurukshetra and Ambala, in contrast, boast more natural wetlands associated with river systems and low-lying areas, supporting a wider range of aquatic and terrestrial species.

The conservation and management of Haryana's wetlands are critical for maintaining ecological integrity and fostering sustainable development. Despite their importance, these ecosystems face numerous threats, including habitat loss, pollution, encroachment, and climate change. Urbanization and infrastructure development have led to the conversion of wetlands for residential, industrial, and commercial purposes, resulting in a loss of valuable habitat and ecosystem services. Pollution from agricultural runoff, industrial discharge, and domestic sewage further degrades water quality and threatens the health of these ecosystems.

Conservation and restoration efforts in Haryana involve a multi-pronged approach, encompassing regulatory measures, community engagement, and scientific research. The state government has enacted laws and policies like the Wetlands (Conservation and Management) Rules, 2017, to regulate activities that may adversely impact wetland ecosystems. Additionally, initiatives like the National Wetland Conservation Program and the Integrated Watershed Management Program provide funding, capacity building, and technical assistance to support wetland conservation efforts. Community participation and stakeholder engagement are crucial for the success of these initiatives. Local communities, including farmers, fishermen, and residents living near wetlands, play a key role in monitoring, managing, and restoring these ecosystems. By involving communities in decision-making processes and raising awareness about wetland importance, conservation efforts can be more effective and sustainable in the long term. Furthermore, scientific research and monitoring are essential for understanding the ecological dynamics of wetlands and identifying priority areas for conservation and restoration.

7.3 POTENTIAL FOR FISHERIES IN VILLAGE PONDS

Village ponds, also known as farm ponds or community ponds, are ubiquitous features of rural landscapes across the globe. These small water bodies serve a multitude of purposes, functioning as vital sources of irrigation water for agriculture, watering holes for livestock, and crucially, as potential hotspots for sustainable aquaculture development. Their size and shape vary considerably, ranging from modest earthen ponds to larger, more intricate reservoirs. Effective management and development of village ponds for fisheries can significantly contribute to improved livelihoods, nutritional security, and overall well-being of rural communities. Village ponds represent a cornerstone for rural development. They provide much-needed water for agricultural activities, particularly during dry seasons when water scarcity is a major constraint. Beyond irrigation, these ponds support livestock rearing by ensuring a reliable water supply for animals. Additionally, village ponds often serve as vibrant community spaces, fostering social cohesion and strengthening bonds through gatherings and events. This inherent multi-functionality makes them integral components of rural life and production systems.

Maximizing the potential of village ponds for fisheries development necessitates effective management practices. A range of activities are crucial, including water quality monitoring, strategic stocking of fish species, habitat enhancement measures, and robust disease prevention protocols. Regular water quality monitoring is paramount to ensure optimal conditions for fish growth and reproduction. Factors like pH, temperature, dissolved oxygen, and nutrient levels all significantly influence the health and productivity of fish populations. Through consistent monitoring, potential issues can be identified early, allowing for timely interventions to mitigate negative impacts. Stocking fish species suited to local environmental conditions, market demands, and cultural preferences is a cornerstone of successful village pond aquaculture. Commonly cultured species include carps like common carp, rohu, and catla, alongside tilapia, pangasius, and freshwater prawns. Stocking densities require careful management to prevent overcrowding and ensure optimal growth rates. Furthermore, integrating multiple species in polyculture systems can enhance overall productivity and bolster resilience against environmental fluctuations.

Creating a suitable environment for fish reproduction and growth is critical, and habitat enhancement measures play a vital role. Providing adequate aquatic vegetation, shelter structures, and appropriate substrate materials fosters a diverse and balanced ecosystem within the pond. Aquatic plants not only offer refuge for fish but also contribute to improved water quality by absorbing excess nutrients and minimizing algal blooms. Shelter structures like floating rafts or submerged logs provide additional hiding places, encouraging natural foraging behavior. Moreover, various substrate materials like gravel or sand offer spawning grounds for specific fish species, facilitating successful reproduction.

Disease prevention is another crucial aspect of pond management, safeguarding the health and productivity of fish populations. Diseases can spread rapidly in confined aquaculture systems, leading to substantial economic losses. Implementing biosecurity measures is essential, including screening incoming fish for diseases, disinfecting equipment, and controlling the movement of personnel and equipment. Regular health monitoring and prompt response to outbreaks are critical for containing diseases and minimizing their impact on fish production.

The potential for income generation and poverty alleviation through village pond fisheries is vast. Fish farming offers a valuable source of protein and essential nutrients, contributing significantly to food security and improved nutrition in rural communities. For smallholder farmers, fish farming in village ponds can generate additional income, diversifying their livelihoods and reducing dependence on traditional crops. Furthermore, the sale of fish and fish products creates opportunities for entrepreneurship and small-scale businesses, stimulating local economies and generating employment opportunities. Fisheries development in village ponds can also have positive environmental implications by promoting sustainable land and water management practices. Integrating fish farming with agriculture, known as integrated aquaculture-agriculture (IAA), can enhance resource use efficiency and reduce environmental degradation. For example, utilizing pond effluents as fertilizer for crops improves soil fertility while reducing the need for chemical fertilizers. Similarly, utilizing agricultural by-products like crop residues and animal manure as fish feed reduces waste and enhances nutrient cycling within agroecosystems.

Beyond economic and environmental benefits, village pond fisheries contribute to social empowerment and gender equality. Women play a significant role in fish farming activities, including processing, marketing, and trading fish products. By actively involving women in fisheries-related enterprises, rural communities can increase household income and empower women economically. Moreover, improved access to fish and fish products enhances dietary diversity and nutrition, particularly for women and children who are often more vulnerable to food insecurity and malnutrition. Government support and investment are essential for unlocking the full potential of village pond fisheries. Policies that promote sustainable aquaculture practices, provide access to credit and technical assistance, and strengthen market linkages are crucial for supporting small-scale fish farmers. Extension services and training programs can help build the capacity of rural communities to effectively manage and operate fish farms. Furthermore, investments in infrastructure development, such as pond construction, water management systems, and transportation facilities, can facilitate the development of fish farming enterprises and enhance market access for rural producers.

7.4 CASE STUDIES

Case studies of flourishing fisheries projects in village ponds provide a rich tapestry of knowledge, revealing effective strategies for sustainable aquaculture development, valuable lessons learned, and replicable best practices. These village ponds serve as crucial sources of fish production, bolstering livelihoods, nutrition, and economic opportunities for local communities. By delving into specific case studies, we can illuminate key factors that contribute to project success, unveil challenges encountered, and strategize methods for overcoming obstacles.

A compelling case emerges from Bangladesh, where community-based fish culture in village ponds has become widely adopted as a tool for poverty alleviation and food security. In many Bangladeshi rural areas, smallholder farmers engage in integrated farming systems, where fish culture in village ponds is interwoven with rice cultivation and livestock rearing. These integrated systems capitalize on the synergistic relationships between various agricultural activities, maximizing resource use efficiency and propelling overall productivity. Projects in Bangladesh that have achieved success have emphasized community participation, capacity building, and the adoption of improved management practices. By engaging local stakeholders in decision-making processes and providing training on pond management techniques, these projects empower communities to take ownership of their fishery resources and cultivate improved livelihoods.

Lessons gleaned from successful fisheries projects in village ponds highlight the paramount importance of stakeholder engagement and participation throughout project design and implementation. Involving local communities in decision-making fosters a sense of ownership and accountability, leading to a greater commitment to project outcomes and their long-term sustainability. Furthermore, capacity building activities, such as training in pond management techniques, fish stocking practices, and marketing strategies, are essential for equipping community members with the knowledge and skills necessary for effective management of their fishery resources. By investing in human capital development, fisheries projects empower communities to achieve self-reliance and become more resilient in the face of environmental and socio-economic challenges.

Best practices gleaned from successful fisheries projects in village ponds encompass the promotion of integrated farming systems, the adoption of low-input and sustainable aquaculture practices, and the implementation of innovative technologies. Integrated farming systems, which combine fish culture with other agricultural activities like rice cultivation, vegetable gardening, and livestock rearing, offer a multitude of benefits, including increased productivity, enhanced resource use efficiency, and income diversification. By harnessing the complementary relationships between different farming components, integrated systems can bolster overall resilience and sustainability, particularly in environments with limited resources.

Low-input and sustainable aquaculture practices, such as the use of locally available feed ingredients, organic fertilizers, and natural pond ecosystems, help minimize environmental impacts and reduce production costs. By promoting ecologically sound approaches to fish culture, fisheries projects can contribute to the conservation of natural resources and the preservation of biodiversity. Furthermore, the adoption of innovative technologies, such as solar-powered aerators, water quality monitoring devices, and mobile-based extension services, can improve the efficiency, productivity, and profitability of fishery operations. By leveraging technological innovations, fisheries projects can overcome logistical constraints, enhance communication and knowledge dissemination, and empower communities to adapt to changing environmental conditions.

Another case study from India showcases the success of community-based fisheries management initiatives in village ponds. In many parts of rural India, community-based organizations, self-help groups, and cooperatives play a pivotal role in managing local fishery resources. These grassroots organizations mobilize community members, organize training programs, and facilitate access to credit, inputs, and markets. By fostering collective action and social cohesion, community-based fisheries management initiatives empower marginalized groups, including women and small-scale fishers, to participate in decision-making processes and reap the benefits of fishery development activities. Furthermore, by promoting sustainable fishing practices, community-based organizations contribute to the conservation of fish stocks and the preservation of ecosystem integrity.

Challenges encountered in implementing fisheries projects in village ponds include limited access to credit and inputs, a lack of technical expertise and extension services, and inadequate infrastructure and market linkages. In many rural areas, smallholder farmers face significant barriers to accessing finance, inputs, and technology, which constrain their ability to invest in fishery development activities. Furthermore, the lack of extension services and technical support hinders the adoption of improved management practices and innovative technologies. Moreover, inadequate infrastructure, such as roads, cold storage facilities, and market centers, limits the marketability of fish products and reduces profitability for both fishers and traders.

To address these challenges, successful fisheries projects in village ponds emphasize the importance of multi-stakeholder partnerships, institutional capacity building, and policy support. By forging partnerships between government agencies, non-governmental organizations, research institutions, and private sector actors, fisheries projects can leverage complementary expertise, resources, and networks to achieve shared goals. Furthermore, by strengthening the capacity of local institutions, such as fisher cooperatives, self-help groups, and extension services, projects can enhance governance structures, promote inclusive decision-making, and facilitate knowledge sharing and technology transfer. Additionally, by advocating for supportive policies and regulatory frameworks, fisheries projects can create an enabling environment for sustainable aquaculture

7.5 SUMMARY

Wetlands, with their diverse habitats and ecological functions, are invaluable ecosystems across the globe. In Haryana, a state in northern India, these ecosystems hold particular significance due to their biodiversity and potential for supporting sustainable fisheries. Recognizing the importance of wetlands and harnessing their fisheries potential can contribute significantly to both environmental conservation and socioeconomic development in the region. Encompassing marshes, swamps, and floodplains, wetlands are characterized by the presence of seasonal or permanent water. They serve as vital breeding grounds, feeding areas, and habitat for a wide range of flora and fauna. Wetlands also provide essential ecosystem services such as water purification, flood control, and carbon sequestration, benefiting both humans and wildlife. In Haryana, wetlands play a crucial role in mitigating monsoon season floods by absorbing excess water and reducing inundation risks in low-lying areas.

The rich biodiversity of wetlands reflects their ecological importance. These ecosystems support a variety of flora and fauna adapted to their unique hydrological conditions. Aquatic plants like reeds, rushes, and water lilies thrive in shallow waters, providing habitat and food for diverse invertebrates, fish, amphibians, and waterfowl. Birds, in particular, are abundant, with many species relying on wetlands for nesting, foraging, and migration. Haryana's wetlands are home to a variety of resident and migratory bird species, including herons, egrets, ducks, and geese, making them important sites for birdwatching and ecotourism.

Conservation of wetlands is critical for preserving biodiversity and maintaining the ecosystem services they provide. However, wetlands in Haryana and other parts of India face numerous threats, including habitat loss, pollution, encroachment, and unsustainable resource exploitation. Conversion of wetlands for agriculture, urbanization, and infrastructure development has led to the degradation and loss of these valuable ecosystems. Pollution from agricultural runoff, industrial discharge, and domestic waste further degrades water quality and threatens the health of wetland habitats. Additionally, overexploitation of wetland resources through fishing and hunting can deplete fish and wildlife populations, disrupting ecological balance and reducing the resilience of wetland ecosystems.

Conservation and restoration efforts are essential for safeguarding Haryana's wetlands and the biodiversity they harbor. These initiatives may encompass habitat restoration, pollution control, community engagement, and policy interventions aimed at sustainable wetland resource protection and management. Habitat restoration projects, such as reforestation, wetland reclamation, and invasive species management, can help rehabilitate degraded wetland habitats and enhance their ecological functions. Pollution control measures, including wastewater treatment, solid waste management, and regulatory enforcement, are necessary to improve water quality and reduce the impact of pollutants on wetland ecosystems.

Community engagement plays a vital role in fostering local stewardship and sustainable resource management. Involving local communities in conservation planning and decision-making processes can raise awareness, foster cooperation, and empower stakeholders to take ownership of wetland resources. Community-based initiatives such as wetland monitoring, ecotourism development, and alternative livelihood programs can provide economic incentives for conservation while promoting sustainable use of wetland resources. Furthermore, partnerships between government agencies, non-governmental organizations (NGOs), academic institutions, and local communities are essential for coordinating conservation efforts, leveraging resources, and sharing knowledge and expertise.

Policy interventions are also critical for effective wetland management and protection in Haryana. Legislation and regulatory frameworks like the Wetlands (Conservation and Management) Rules, 2017, provide legal mechanisms for identifying, delineating, and protecting wetland areas. These rules require states to develop wetland inventories, management plans, and conservation strategies to ensure sustainable wetland resource management. Financial incentives such as grants, subsidies, and tax breaks can encourage landowners and stakeholders to participate in wetland conservation and restoration activities. Strengthening enforcement mechanisms, capacity building, and public awareness campaigns are also crucial for ensuring compliance with wetland regulations and promoting a culture of conservation among stakeholders.

Harnessing the potential of wetlands for fisheries presents valuable economic opportunities while promoting sustainable resource management and environmental conservation. Wetlands support diverse fish species adapted to a range of aquatic habitats, from rivers and lakes to ponds and marshes. These ecosystems provide essential spawning, nursery, and feeding grounds for fish, contributing to the productivity and resilience of freshwater ecosystems. In Haryana, wetlands offer significant potential for fisheries development, particularly in rural areas where fishing provides livelihoods for thousands of people.

Expanding sustainable aquaculture and fisheries practices in wetlands can contribute to food security, poverty alleviation, and rural development in Haryana. Aquaculture, or fish farming, involves cultivating fish in controlled environments like ponds, tanks, and cages. In wetlands, aquaculture can be integrated with traditional farming practices like rice cultivation to maximize land use and productivity. Fish farming provides employment opportunities for small-scale farmers and fishers, particularly women and marginalized communities, who often rely on fisheries for their livelihoods.

7.6 CONCLUSION

Haryana's fisheries sector faces a confluence of challenges and opportunities, mirroring the global discourse on sustainable fisheries management and development. As a landlocked state with limited natural water resources, Haryana encounters unique constraints in expanding its fisheries. However, the state has demonstrated promising potential for aquaculture development through the adoption of modern technologies, innovative practices, and supportive policies. The success of Haryana's fisheries hinges on a multifaceted approach that addresses social, economic, and environmental considerations, while promoting sustainable development and responsible stewardship of aquatic resources.

Looking ahead, considerations of resilience, adaptation, and collaboration are paramount for Haryana's fisheries. In the face of climate change, urbanization, and other mounting pressures, fisheries stakeholders must demonstrate adaptability and resilience, embracing new technologies and practices to navigate uncertain waters. Collaboration between government agencies, research institutions, industry players, and local communities is essential to foster innovation, share knowledge, and coordinate efforts towards sustainable fisheries management. By working together, stakeholders can address shared challenges, leverage collective expertise, and maximize the benefits of fisheries for current and future generations.

Recommendations for sustainable development in Haryana's fisheries sector encompass a range of strategies designed to balance economic growth with environmental conservation and social equity. First and foremost, robust data collection, monitoring, and research are crucial for informing evidence-based decision-making and ensuring the sustainable use of aquatic resources. This entails conducting comprehensive assessments of fish stocks, habitat health, and socio-economic dynamics to identify areas of concern and opportunities for improvement. Investing in scientific research, capacity building, and technology transfer will bolster the knowledge base and technical skills necessary for sustainable fisheries management.

Furthermore, strengthening regulatory frameworks and enforcement mechanisms is critical to safeguard fisheries resources and promote responsible practices. This necessitates implementing and enforcing fisheries laws, regulations, and licensing requirements to deter overfishing, habitat degradation, and illegal activities. Enhancing compliance monitoring, surveillance, and enforcement capacity will deter illegal fishing practices and foster a culture of accountability among stakeholders. Additionally, promoting community-based management approaches, such as co-management and participatory decision-making, can empower local communities to take ownership of their fisheries resources and contribute to their sustainable management.

Promoting sustainable aquaculture practices is another key priority for the future of Haryana's fisheries. This necessitates the adoption of environmentally-friendly and socially-responsible aquaculture techniques, such as integrated multi-trophic aquaculture (IMTA), recirculating aquaculture systems (RAS), and organic farming methods. By minimizing environmental impacts, conserving water resources, and reducing reliance on wild-caught fish for feed, sustainable aquaculture practices can alleviate pressure on natural ecosystems while meeting the growing demand for fish protein. Investment in research and extension services to disseminate best practices and provide technical support to fish farmers will facilitate the transition to more sustainable aquaculture systems.

Moreover, enhancing market access and value chain development is essential for promoting economic growth and livelihood opportunities in Haryana's fisheries sector. This includes improving infrastructure, transportation, and cold chain facilities to reduce post-harvest losses and ensure the quality and safety of fish products. Strengthening linkages between fish producers, processors, retailers, and consumers through market-based incentives, value-added processing, and branding initiatives can enhance the competitiveness of Haryana's fisheries products in domestic and international markets. Additionally, promoting fair trade practices, equitable distribution of benefits, and inclusive value chains will ensure that fisheries contribute to poverty alleviation and social development in rural communities.

Finally, promoting environmental sustainability and biodiversity conservation is paramount for the long-term viability of fisheries in Haryana. This necessitates protecting and restoring critical habitats, such as wetlands, rivers, and lakes, that provide essential ecosystem services and support fisheries productivity. Implementing habitat restoration projects, pollution control measures, and watershed management initiatives will improve water quality, enhance habitat connectivity, and mitigate the impacts of climate change on aquatic ecosystems. Furthermore, promoting sustainable fishing practices, such as selective gear use, habitat-friendly fishing techniques, and seasonal closures, will help maintain healthy fish populations and ensure the resilience of aquatic ecosystems in the face of environmental change.

REFERENCES

1. Dey, M. M., & Ahmed, M. (2005). Aquaculture—Food and Livelihoods for the Poor in Asia: A Brief Overview of the Issues. Aquaculture Economics & Management, 9(1-2), 5-20.

2. Kumar, P., & Sen, S. (2006). Fisheries Development in India: The Political Economy of Unsustainable. Economic and Political Weekly, 41(26), 2767-2777.

3. Jhingran, V. G. (1991). Fish and Fisheries of India. Hindustan Publishing Corporation.

4. FAO. (2018). The State of World Fisheries and Aquaculture 2018: Meeting the Sustainable Development Goals. Food and Agriculture Organization of the United Nations.

5. World Bank. (2013). Fish to 2030: Prospects for Fisheries and Aquaculture. World Bank Report Number 83177-GLB.

6. Naylor, R. L., Goldburg, R. J., Primavera, J. H., Kautsky, N., Beveridge, M. C., Clay, J., ... & Troell, M. (2000). Effect of Aquaculture on World Fish Supplies. Nature, 405(6790), 1017-1024.

7. Bostock, J., McAndrew, B., Richards, R., Jauncey, K., Telfer, T., Lorenzen, K., ... & Corner, R. (2010). Aquaculture: Global Status and Trends. Philosophical Transactions of the Royal Society B: Biological Sciences, 365(1554), 2897-2912. https://doi.org/10. 1098/rstb.2010.0170

8. De Silva, S. S., & Davy, F. B. (2010). Aquaculture Successes in Asia: Contributing to Sustained Development and Poverty Alleviation. FAO Fisheries and Aquaculture Technical Paper, No. 500.

9. Tacon, A. G. J., & Metian, M. (2013). Fish Matters: Importance of Aquatic Foods in Human Nutrition and Global Food Supply. Reviews in Fisheries Science, 21(1), 22-38.

10. Directorate of Fisheries, Government of Haryana. (2020). Annual Report 2019-20. Government of Haryana.

11. Pillay, T. V. R. (1990). Aquaculture: Principles and Practices. Blackwell Publishing.

12. Ahmed, N., & Thompson, S. (2019). The Blue Dimensions of Aquaculture: A Global Synthesis. World Development, 120, 65-77.

13. Beveridge, M. C., Thilsted, S. H., Phillips, M. J., Metian, M., Troell, M., & Hall, S. J. (2013). Meeting the Food and Nutrition Needs of the Poor: The Role of Fish and the Opportunities and Challenges Emerging from the Rise of Aquaculture. Journal of Fish Biology, 83(4), 1067-1084.

14. Central Institute of Freshwater Aquaculture (CIFA). (2015). Handbook of Fisheries and Aquaculture. Indian Council of Agricultural Research.

15. Gupta, M. V., & Acosta, B. O. (2004). A Review of Global Tilapia Farming Practices. Aquaculture Asia, 9(2), 7-12.

16. Indian Council of Agricultural Research (ICAR). (2013). Handbook of Fisheries and Aquaculture. ICAR.

17. Jhingran, A. G. (1982). Fish and Fisheries of India. Hindustan Publishing Corporation.

18. Kumar, B. G., & Dey, M. M. (2006). Integrated Fish Farming: A Blueprint for Sustainable Agriculture. Aquaculture Asia, 11(3), 19-25.

19. National Fisheries Development Board (NFDB). (2018). Fish Production in India 2017-18. NFDB.

20. NBFGR. (2017). Status of Freshwater Fish Diversity in India. National Bureau of Fish Genetic Resources, ICAR.